'In a stimulating dialogue between theory an
Innovation explores how digitally-enabled process
services and products that change the ways we w
a broad knowledge perspective and grounded in a number of interesting real life
cases, the book offers students and managers a rich conceptual understanding
and critical insights into the complexities and opportunities of innovation. This
is an important, approachable contribution elaborating digital as well as sustainable prospects for the future.'

—*Professor Birgit Helene Jevnaker, BI Norwegian Business School, Norway*

'This textbook offers a timely and thorough introduction to the fascinating
world of digital innovation. The authors masterfully combine attention for
recent developments in practice with critical reflection rooted in theories of
knowing, organizing, and innovating. The book steers free from hype and gloom
to offer in-depth understanding instead.'

—*Professor Hans Berends, Vrije Universiteit Amsterdam, the Netherlands*

'This is a soundly-researched and well-presented text. It will be welcomed by
senior students studying digital innovation, especially with reference to the crucial role of knowledge in supporting and inspiring the organisational development of new processes, products and services in the second decade of the
twenty-first century. The integration of underpinning theory, case study examples, key concepts, and questions for discussion, is bound to earn a place for this
book on many university reading lists.'

—*Professor Hazel Hall, Edinburgh Napier University, UK*

'Digital innovation stands out as one of the key factors in the vital and exciting
pursuit of many firms to improve their social, and economic performance. In their
timely book, Newell, Morton, Marabelli, and Galliers offer a comprehensive view
and critical thinking in relation to this significant theme. An important textbook
that will become an invaluable resource to both students and (future) managers.'

—*Dr. Robert Verburg, Delft University of Technology, the Netherlands*

'This is a must-read book for students – as current or future managers – which
aims to understand the often complex and messy processes surrounding digital
innovation. It offers the fundamentals of responsibly managing digital innovation from a knowledge perspective. In exploring the notion of organising for
digital innovation, the authors focus on sources of knowledge and how knowledge is acquired and shared in innovation processes and how this is influenced
by organisational structures and processes. The novelty of this book is that it
stresses the concepts of responsible and frugal innovation and their relevance to
digital innovation. In fact, this book presents a warm invitation to reflect on the

purpose of digital innovation in whatever type of organizational context. Making a profit should not prevent those responsible for decision making from also considering the environmental and social consequences of the innovation.'

—*Silvia Gherardi, University of Trento, Italy*

'This book introduces key aspects of digital innovation and digitally enabled innovation processes in an excellent way, mixing interesting case studies with explanations of concepts and insights from the research literature. In addition to the wide range of topics pertaining to digitalization in a narrow sense the authors have also included important wider references to the way people and digital technologies interact and the intended and unintended consequences of such interactions.'

—*Professor Terje Grønning, University of Oslo, Norway*

'This excellent textbook fills a gap in the market. Its focus on digital innovation and its in-depth exploration of the topic constitute a timely and original addition to the literature on innovation. The book provides readers with a solid theoretical and empirical background for understanding and managing the opportunities and challenges of the digital transformation of the economy, organizational structures and business models. This is aided by the use of real-world case studies which open each chapter.'

—*Professor Klaus Nielsen, Birkbeck University of London, UK*

MANAGING DIGITAL INNOVATION

A KNOWLEDGE PERSPECTIVE

SUE NEWELL, JOSH MORTON,

MARCO MARABELLI, AND

ROBERT D. GALLIERS

First published 2020 by
RED GLOBE PRESS

Red Globe Press in the UK is an imprint of Springer Nature Limited, registered in England, company number 785998, of 4 Crinan Street, London, N1 9XW.

Red Globe Press® is a registered trademark in the United States, the United Kingdom, Europe and other countries.

ISBN 978-1-137-43429-6 paperback

https://doi.org/10.26777/978-1-137-43240-7

CONTENTS

List of Figures   x
List of Tables   xiii
Preface   xiv

1   The Changing Context of Work: Implications
    for Knowledge and Innovation   1
    Introduction: The rise of the knowledge economy   2
    The growing importance of digital innovation   5
    The birth of knowledge management   8
    Knowledge as a concept   11
    Possessed and practised knowledge   13
    Conclusions   17
    Additional suggested readings   17
    References   18

2   The Innovation Process   21
    Case: Enterprise system implementation at Tech Co.   22
    Introduction: Rogers's diffusion of innovation   25
    Innovation: An overview   26
    Rogers's model of the diffusion of innovation   33
    An interactive view of innovation   37
    Absorptive capacity   40
    Power and knowledge absorption   44
    Conclusions   45
    Additional suggested readings   46
    References   46

3   Organising for Digital Innovation   49
    Case: CommCo – The use of social media to encourage
    simultaneously *univocality* and *multivocality*   50
    Managing different types of content on the intranet   50
    Nurturing a communication culture for OPC and UGC   52
    Introduction: Classic organisation design   53
    Dysfunctions of bureaucracy   55
    Contingency theories   58

Bureaucracies: Digital innovation and change 60
New forms of organising 62
Aligning old and new organisational forms 70
Conclusions 71
Additional suggested readings 73
References 73

4   **Strategising for Digital Innovation** 77
Case: ChinaTicketCo – event ticketing in China 78
Phase one: Ticketing as a transactional practice 78
Phase two: Ticketing as a transactional and a
relational practice 79
Phase three: Ticketing as a transactional, relational
and experiential practice 80
Introduction: Strategy versus strategising 81
Strategy as a deliberate planning process 82
Porter's three strategies 86
An alternative, emergent view of strategy 89
Strategy-as-practice and ambidexterity 91
Digital technology and strategy-as-practice 94
Conclusions 97
Additional suggested readings 98
References 99

5   **Projects and Teaming** 101
Case: SkinTech 102
Introduction: Routines and innovation 105
Projects, teams and teaming 108
Project management 110
Types of projects and project management 116
Sources of uncertainty in projects and project learning 119
The nested nature of projects: Projects in their
institutional context 124
Conclusions 127
Additional suggested readings 128
References 128

6   **Project Liminality and Open Innovation** 131
Case: Project liminality and open innovation – Defence-co's
InnovationJam and the city of Vienna's co-creation
project 132

Initiative #1: Defence-co's InnovationJam   132
Initiative #2: The city of Vienna's co-creation project   135
Introduction   137
Project organising and liminality   137
Paradoxes in projects and liminality   141
Open innovation   144
Who is involved in open innovation?   145
Inside-out and outside-in open innovation   148
Business models and innovation   149
Open innovation and services   151
Open innovation and networking: Brokering
and boundary spanning   153
Open innovation: The search issue   155
Open innovation and governance   157
Conclusions   158
Additional suggested readings   159
References   160

7   **The Role of Objects in Organising for Digital
Innovation**   **163**
Canada-Care case: The role of objects in an
innovative healthcare initiative in Canada   164
Objects used in the initiative   166
Summary   169
Introduction: Agency, objects and organising   169
Objects and innovation processes   171
Strategic objects   174
Boundary objects   177
The innovative power of objects   181
Human engagement in objects: The role of emotions   185
Conclusions   186
Additional suggested readings   188
References   188

8   **Explicit Digital Connectivity, Knowledge
and Innovation**   **191**
Case: Digital innovation in China (Taobao
Villages)   192
Introduction   194
Implicit versus explicit digital connectivity   197

CONTENTS

Explicit digital connectivity and e-Commerce   198
Explicit digital connectivity and use of social media
within organisations   201
Explicit digital connectivity and crowdsourcing   204
Peer-to-Peer crowd involvement: The sharing economy   209
Explicit digital connectivity and work–life boundaries   212
Conclusions   216
Additional suggested readings   217
References   217

9   Opportunities and Challenges for Innovation
    Related to Implicit Digital Connectivity   223
    Case: Sensors in the automotive industry   224
    Introduction: Data and the digital world   229
    Algorithmic decision-making and big and little data   232
    Implicit digital connectivity and computing applications 236
    Implicit digital connectivity and digitised objects   241
    Innovation opportunities from big and little data   246
    Making digital innovation more ethical   247
    Conclusion   249
    Additional suggested readings   251
    References   251

10 The Future of Digital Innovation: The Role
    of Responsible and Frugal Innovation   255
    Case: Frugal innovation in India   256
    The Tata Nano   257
    The Jaipur Foot   258
    Introduction   260
    Responsible innovation: What is it and why does it matter? 262
    Key opportunities and challenges for responsible
    digital innovation   266
    Frugal innovation: An emerging consideration
    in developing and developed economies   268
    Frugal innovation and established models of innovation 269
    The challenges of frugal innovation   273
    Key actions for successful responsible and
    frugal innovation   275
    A summary of responsible and frugal innovation:
    Key areas of applicability   278

Conclusions 281
Additional suggested readings 282
References 282

Notes **284**
Index **286**

LIST OF FIGURES

1.1 Developing and implementing new ideas 7
1.2 Knowledge management in organisations 11
2.1 Enterprise systems in organisations 25
2.2 Tracking transportation of goods 28
2.3 Modern laser disk (CD) reader 30
2.4 Linear and interactive views of innovation 38
2.5 Key steps in absorptive capacity 40
2.6 Example process of absorptive capacity 42
2.7 An interactive view of absorptive capacity 43
3.1 Servers handling an organisation's
digital communication 51
3.2 Authority is a multifaceted construct 54
3.3 Different ways to view how organisational
components can be combined 59
3.4 Centralisation versus decentralisation 61
3.5 Hierarchical versus 'flat' organisations 62
4.1 New ways to sell, through online platforms 79
4.2 Porter's five forces 82
4.3 Conception and execution of a strategy 87
4.4 Porter's three generic strategies 89
4.5 Practice, praxis and knowledge exploration
and exploitation 92
4.6 New technologies such as cloud computing
enable emerging innovation 96
5.1 Clinical research and innovation 105
5.2 Grouping is key for facing team-based challenges 110
5.3 A word cloud showing where emphasis in project
management typically lies 113
5.4 An example of a PERT chart in the software
WBS Schedule Pro (screenshot reproduced
with permission from WBS) 114
5.5 Development of innovative drugs 118
5.6 The long road to a new medicine 121
5.7 Challenges of software innovation 125

6.1 What is a Jam? Poster used to advertise the Defence-co
InnovationJam (poster reproduced with permission
from designer) 134

6.2 Overview of the city of Vienna's co-creation project 136

6.3 Liminality represents a 'passage'
from one condition/state to another 140

6.4 In projects, a sense of community
within a liminal space might be detrimental
to those outside a community 142

6.5 Kodak was not able to understand long-term
challenges of being innovative 146

6.6 Business models are relevant to execute
innovative strategies 150

6.7 Lego's (open) innovative approach to services 152

6.8 A diagram representing a broker
between two communities 155

6.9 Overview of stakeholders and their involvement
in the city of Vienna's co-creation project 155

7.1 At Canada-Care, the clinician's engagement was
paramount for improving healthcare service delivery 165

7.2 The medical sheet used at Canada-Care 168

7.3 Presentations, for instance, included objects
such as PowerPoints, which were proven helpful
to translate knowledge across practitioners 170

7.4 Strategic tools such as Gantt and Pert charts
are used in most business meetings 175

7.5 At the Berkeley Museum of Vertebrate Zoology
in California, various specimens promoted
knowledge-sharing between scientists 179

7.6 Imbrexes and tegulae are interlocked,
yet they are two distinct entities 183

8.1 Smartphone apps allow the sharing
of our whereabouts constantly 197

8.2 eBay and similar websites connect people
from all over the world 199

8.3 e-commerce platform 200

8.4 It is often difficult to find the right balance between
using technology intensively and responsibly 206

8.5 Millions of people daily post online reviews
to various websites 209

8.6 We cannot ignore the dark side of innovation uses 210

8.7 The sharing economy has reshaped many industries
worldwide, the most famous probably being the taxi industry 211

8.8 Working from home is more and more common,
yet it also poses social issues   213

9.1 Generali's PAYD model, technology and processes   226

9.2 Progressive's PAYD model, technology and processes   228

9.3 Expanded computer storage is one of the factors
enabling datafication   230

9.4 Algorithms enable decision-making capabilities   233

9.5 Onboard computers are able to capture the minutiae
of our life when we drive our car   235

9.6 Search engines can drive innovative discoveries   237

9.7 Surveillance cameras contribute to address crimes,
yet they pose societal issues   242

9.8 Monitoring workers' whereabouts might increase
short-term efficiency while over-stressing employees   243

9.9 Employee tracking systems   244

9.10 The Dark Web and the underground internet
challenge current laws and regulations   248

10.1 Frugal innovations in developing countries help
in exploiting existing resources with a view
to creating new sources of wellbeing   261

10.2 India and China are among these countries
where frugal innovation has found fertile ground   263

10.3 Nuclear plants, and associated innovation,
have been hugely regulated in the past decades   268

10.4 Innovation initiatives in Silicon Valley include
those addressing sustainability   271

10.5 Creativity is key to frugal innovation   275

LIST OF TABLES

2.1 Types of innovation 30

2.2 Firm/user perspectives of innovation 32

3.1 An example of simple modularisation 66

3.2 A more complex example of modularisation 66

6.1 Differing assumptions of closed
and open innovation approaches 145

6.2 Selecting different open innovation strategies
(adapted from Felin and Zenger, 2014) 158

7.1 The power and role of objects 184

8.1 Critical actors in the rural e-commerce ecosystem 195

8.2 Orchestrated and organic approaches
(from Cui et al., 2017) 202

PREFACE

Digital innovation is often seen as the new normal these days, but digital innovation requires careful treatment and should not be assumed to automatically arise from the ever-increasing power and capability of digital technologies. Though potentially transformative, digital innovation is complex and fraught with difficulty. This book aims to unpack the issues and provide guidance to the unsuspecting. Beware the Siren calls that accompany many of the latest popular texts on the subject.

Gordon Moore observed in 1965 that the number of transistors on an integrated circuit doubles every two years. This became known as Moore's law and underlies the many advancements that have been made (and continue to be made) in digital electronics since. Prices of microprocessors have fallen sharply while memory capacity has increased exponentially, leading to the ability of organisations to do a lot more with digital technology than ever before. This has spurred a massive rise in digital innovations that is ongoing and potentially transformational for organisations, irrespective of their industry – for their products and services and for the ways in which they operate. Moreover, digital innovations have changed how organisations 'do' innovation, for example, in enabling organisations to include people from across the globe in generating ideas for new products and services or in coming up with solutions to problems based on crowdsourcing using platform-based social software applications.

Examining these digitally enabled innovation processes and the products and services that have been generated is the focus of this book. Importantly, in examining these processes we explore not only the positive consequences for organisations but also the negative and often unintended consequences – for the organisations themselves (as when an innovation project does not add the value expected) and for employees, customers, supply chains and, more generally, for societies and the environment.

How the textbook is structured

The purpose of this book is to provide students – as current or future managers – with an understanding of the often complex and messy processes surrounding digital innovation. Digital innovation processes are difficult to fully plan for in advance, so managers need to constantly revisit what is being done and what the likely consequences may be. Technology does not produce simple causal outcomes; rather, technology interacts with people and existing social structures so that the same technology can have different consequences in different places and at different times. Understanding innovation processes and the role of digital technology as both an input to and an output of these processes can help managers make better decisions – whether these decisions relate to how to manage the innovation process itself (including how to use digital technology to support the innovation process) or in deciding what digital products and services to attempt to develop (not just in terms of adding financial value to their organisation, but in adding social and environmental value in addition).

In constructing this textbook, we have outlined 10 chapters which together explain the fundamentals of managing digital innovation from a knowledge perspective. Chapters 1 and 2 provide the background context for the book and outline the move from traditional economies (based on land, labour and capital) to knowledge economies. Here we discuss that, today, it is recognised that knowledge is distributed among all employees and that to foster innovation, which is increasingly important, it is necessary to constantly explore and exploit knowledge processes that often involve digital technologies. These chapters also consider what we mean by innovation, particularly digital innovation, and knowledge (seeing knowledge as the ability to draw distinctions in patterns of data) and innovation (the creation and spread of new ideas).

The themes in Chapters 3, 4 and 5 focus on organising for digital innovation, strategising for digital innovation and the importance of projects and teaming in innovation. In exploring the notion of organising for digital innovation, we consider how organisational designs (structure and process) have changed over time, often moving from bureaucratic structures to project-oriented modes of organising that are more flexible and so can support innovation, including digital innovation. We focus in particular on sources of knowledge and how knowledge is acquired and shared in innovation processes and how this is influenced by organisational structures and processes. The importance of how organisations strategise in relation to digital innovation considers that, in practice, strategies often do not work out in the ways anticipated and so we explore strategy literature that recognises the emergent and dynamic nature of strategy. This is most commonly referred to as strategising, which reflects the everyday 'doing' of strategy in organisations. This is considered whilst recognising the importance of strategy and strategising that simultaneously exploits and explores knowledge, considering aspects of digital strategising and its relevance to digital innovation in passing. This provides a basis to examine the prevalence of organising work through projects in today's organisations and some of the reasons for this. We also explicitly consider how innovation projects can be managed when there is uncertainty (including 'unknown-unknowns'), where plans and controls can stifle rather than support innovation in the face of continuously emergent knowledge.

Chapter 6 and 7 begin to explore more deeply the issues around digital innovation and the relevance of a number of specific issues regarding organising, work and knowledge in contemporary organisations and so consider some of the opportunities and challenges for digital innovation. First, in Chapter 6, we look in more detail at how (digital) innovation projects can be managed, whether undertaken within an organisation's boundaries or as a more open process, typically seeking to expand innovation processes to include others outside of the organisation. In terms of internal projects, the concept of liminality is also introduced here to explore the trade-off between being separated from the ongoing workings of an organisation and having to convince the rest of the organisation that the outcome is a good one at the end of the process. This provides a platform to explore, in Chapter 7, the role of material objects in processes involving knowledge creation and sharing and in facilitating digital innovation. Specifically, we explain the role of objects in organising and we discuss the idea that objects (as well as people) have agency (i.e., make things happen) in relation to innovation processes.

We continue these themes in Chapters 8 and 9 and consider explicit and implicit digital connectivity. Chapter 8, focusing on explicit digital connectivity, considers how individuals, organisations and even regions are leveraging the digitally 'connected world' in which we live to support innovation activities. This is not only in developed economies but in less developed economies as well. We illustrate how e-commerce applications generally – and communication applications specifically – can and are being used in many different ways to support innovation in a variety of different contexts. We then shift the focus in Chapter 9 to opportunities and challenges for innovation related to implicit digital connectivity and consider innovative processes underpinning the use of large datasets. This includes personal information to profile individuals, analyse consumer behaviours and create predictive scenarios. The peculiarity of these innovation processes is that data are analysed without the explicit intent of the data creator (e.g., a social media user) or captured without the awareness of the individual (e.g., people 'caught on camera' on CCTV). This therefore contrasts implicit digital connectivity with explicit digital connectivity as we have examined them in these two chapters.

We conclude the book in Chapter 10 with an outlook on the future of innovation, including digital innovation. Here we focus, in particular, on reviewing the concepts of responsible and frugal innovation and their relevance to digital innovation. We consider what motivates the development of these models of innovation and the potential consequences of their use in both developed and developing countries. Responsible innovations are aimed at innovating for the social benefit of individuals and societies, with an emphasis on sustainability, acceptability and social desirability. Frugal innovation has similarities but with more focus on innovating with fewer resources and being creative in developing innovations which have a core purpose of doing more with less and changing society for the better. We offer this final chapter as a way for readers to think seriously about the purpose of innovation, digital or otherwise, in whatever type of organisational context – making a profit can be a purpose we argue, but that should not prevent those responsible for decision-making in also considering the environmental and social consequences of the innovation.

1 THE CHANGING CONTEXT OF WORK: IMPLICATIONS FOR KNOWLEDGE AND INNOVATION

Summary

This chapter considers the move from traditional economies (based on land, labour and capital) to knowledge economies. In traditional economies, knowledge was centralised and most workers were just 'a pair of hands' – labourers. Today, it is recognised that knowledge is distributed among all employees and that to foster innovation, which is increasingly important, it is necessary to constantly explore and exploit knowledge processes that often involve digital technologies. The chapter also considers what we mean by innovation, particularly digital innovation, and knowledge, seeing knowledge as the ability to draw distinctions in patterns of data and innovation as the creation and spread of new ideas.

Learning Objectives

The learning objectives for this chapter are to:

1. Understand the rise of the knowledge economy
2. Recognise the importance of innovation, especially digital innovations, for organisations
3. Appreciate the potential dark side of continuous innovation, especially from an environmental perspective
4. Comprehend the link between innovation and knowledge
5. Be able to define knowledge and appreciate different approaches to knowledge management.

Introduction: The rise of the knowledge economy

The Industrial Revolution was propelled by technological and social innovations and access to what economists describe as the 'traditional factors of production' – land, labour and capital:

- Land was necessary for mining the raw materials required for producing goods or for building factories in which the work of production could be carried out (e.g., land in Lancashire, England was home to over 2,500 cotton mills by the 1860s);

- Labour was needed as the driving force to actually do the work of making goods (e.g., 440,000 labourers were employed in nineteenth-century Lancashire cotton mills, producing half of the world's cotton);

- Capital was needed to develop and buy technology (e.g., the Spinning Jenny, invented in 1764, that mechanised cotton spinning and increased the output of a worker by allowing each labourer to work multiple spools at once) in order to expand the enterprise and so create opportunity for more production (by purchasing additional land and labour), and make more profit (i.e., capital), thereby creating a virtuous circle of ever-increasing economic growth and profits.

Importantly, in this mix, labour is treated as a factor of production, no different from land and capital – a resource that must be acquired and managed, just like any raw material or piece of technological equipment used in production. The 'hands/body' were what mattered, rather than the 'brain', at least for the majority of employees in the early period of industrialisation. Indeed, the whole premise of Taylor's (1911) so-called Scientific Management was that firms were inefficient because employees were using their brains, and they used their brains to ensure that they did the minimum amount of work – described as systematic soldiering. Labourers could slack like this because they, and not their managers, had knowledge about the work they were doing and how much effort they were expending. When a manager asked a person to 'work harder' they could reply that they were 'working as hard as possible' and it was very difficult for the manager to argue because they did not know what they should expect.

Based on this realisation, Taylor's idea was to transfer the workers' knowledge to managers (an entirely different and very small group compared to the basic 'labourer') so that labour could be more efficiently utilised as a resource. Scientific Management was therefore a system designed to study work processes, define the most efficient way to do each task and then pay for output. More specifically, time and motion studies were developed under the guise of Scientific Management where each movement of a particular work task is timed (jobs being broken down into their component tasks rather than studied as a whole), new ways of carrying out each task are experimented with, and again timed, and ultimately the most efficient way of carrying out a task is identified. Time expectations for these newly defined tasks are allocated, and workers are then compared against this standard and paid according to the amount of output they achieve – payment-by-results, instead of pay per hour as was the norm before Taylor's intervention. In this way, labourers became 'cogs in the machine' of a factory in a very real sense.

Automation and the importance of services

What we see from the brief review of the design of work at the turn of the twentieth century is that labour did not include those scientists/inventors who created and developed the breakthrough technologies, nor the managers (or owners) who organised production to generate profits. However, in the post-World War II era, a number of changes occurred that have led to a rethink (in some organisations and places at least) of the way labour is perceived and used. This includes changes driven by technological advancements and the internet, such as offshoring, business model innovation and globalisation of markets. Further, two prominent related issues are automation and the importance of services, which we use as examples to illustrate the rise of the knowledge economy and changes to the view of labour in this chapter:

1. ***Automation:***
    Gradually, new technologies have been developed that take over many of the routine physical tasks that were previously done by manual labourers (e.g., using robotics on assembly lines or mechanised equipment in farming). Moreover, this automation then extended beyond the scope of routine manual jobs and into the office. As a result, with the increasing use of Information Technology (IT), many other jobs have been automated or partially automated (e.g., bank tellers no longer process cheques by hand, computers do computations that were previously done by people). Automation is based on the fact that machines can be more efficient and reliable than people – because, for example, they don't need breaks, can work at a constant (and fast) speed; don't get sick and leave; and, most importantly, are cheaper (if considered over their lifetime output). This is why it came as a great surprise when research began to identify what was called a 'productivity paradox' (Brynjolfsson, 1993) – whereby the 1970s investments in technology (and, in particular, IT used for office automation) did not appear to be related to gains in productivity in the same way as investments in other technologies had been previously (e.g., in relation to factory or farming automation).

    Nevertheless, subsequent analysis demonstrated it was not that this type of computer automation had had no impact, but that the impact of IT is dependent on how it is used to enable, for example, changes in work processes. Therefore, IT does not so much influence the quantity of work undertaken, but rather its quality and the ability to innovate (Hammer,

1990; Brynjolfsson and Hitt, 2000). In other words, IT does not itself *create* productivity gains, but depends on the conditions of use and often it takes time to develop the benefits from IT adoption. This is a theme of this book and one that we take up throughout the different chapters. More generally, the use of technology to automate work indicates that, where people continue to be involved, it is because of the skills and abilities they have which are not easily replaced using technology (at least at the current time). The physical (hands/body) may be replaceable in some jobs, but the ability to be flexible and responsive and to innovate remain, for now, uniquely human skills and abilities (although see Chapter 9, where we discuss how datafication and algorithmic decision-making are changing this).

2. ***Importance of services:***
With more competition from global markets, firms increasingly thrive based not simply on their ability to produce products for which customers are prepared to pay; they also need to offer a range of services that attract and retain customers. As a result, the provision of services has become as important as the production of goods (Oliva and Kallenberg, 2003) even in what appear to be traditional manufacturing or product-based companies. For instance, car dealerships don't just sell cars (products), they sell financial services (selling customers extended warranties and loans or leases so that they can pay for a vehicle), insurance (so that the buyer can insure the purchased vehicle, for example using products such as GAP – Guaranteed Asset Protection – that covers the difference between the remaining balance on a loan and the car's actual value if a vehicle is lost or stolen), and also after-sales services (to keep cars on the road when things go wrong or simply to maintain cars so that they are roadworthy).

A 2012 *Forbes* article entitled 'The Surprising Ways Car Dealers Make The Most Money Off You' demonstrates the importance of these services: for example, finance and insurance products represented about 3 per cent of revenues, but 20 per cent of the gross profits for the Asbury Automotive Group, whereas for the Penske Automotive Group, services and parts represented 13 per cent of annual revenues but 44 per cent of gross profits. While aspects of services can be automated, services are often more people-intensive than manufacturing and don't rely simply on the hands/body of the worker but require, for instance, social skills and the ability to adapt to changing or diverse contexts.

Both automation and the growth in services can be described as innovations – firms have innovated in relation to the organisational processes that support production and have diversified beyond a production mentality to a service culture. Innovation is therefore key to our understanding of the ways organisations have adapted (and are adapting) to their ever-changing context and how innovation requires much more than the hands/body of the worker.

The growing importance of digital innovation

As global competition has increased, a general trend has been the need for companies to increase their pace of innovation (Bolwijn and Kumpe, 1990) – that is, to introduce changes to products or services or new organisational arrangements or business models that support efficient production and the effective development and selling of products and services. These changes often involve digital innovations – in other words, innovations that incorporate IT into products, services, processes and business models to (hopefully) add value for stakeholders. Digital technologies are many and various, but they are all based on using computing power that records, stores and transmits data quickly and accurately. Importantly, this computing power has increased dramatically so that these data-processing capabilities provide myriad opportunities for the development of innovations in, for example:

- New products (e.g., autonomous vehicles that have sensors which collect millions of pieces of data that are analysed based on algorithms to navigate the world around them);

- New services (e.g., online chat applications that connect people around the world through transferring messages using the internet);

- New organisational processes (e.g., use of robotics and artificial intelligence to perform tasks previously done by humans, such as picking orders from a warehouse for an online sales process or reading medical scans to detect cancers);

- New business models (e.g., using big data and analytics to personalise marketing campaigns so that consumers receive more targeted advertising based on their previous online searches and/or purchase histories).

This increase in digital innovation allows products and services to be continuously updated based on data about usage in order to create more sales, based supposedly on improved design or functionality (Apple products are a good example of this). This phenomenon has been described as planned obsolescence, a phrase popularised by Brooks Stevens in 1954, when he used the term to describe instilling in the buyer the desire to own something a little newer, a little better, a little sooner than is necessary.

This is very different from earlier times. Henry Ford, for example, the founder of the Ford Motor Company, was said to have remarked about the Model T (the first car produced on an assembly line) in 1909 that any customer can have a car painted any colour that he wants, so long as it is black. Although the Model T was very much an innovation – it was the first car designed that an 'average' middle-class American could afford – it subsequently lasted much longer than current models of cars, being in production from 1908 to 1927. Indeed, as the above quote testifies, it did not come in multiple variants to appeal to different audiences, and modifications to the design were much slower than is currently the norm in the automotive industry.

The downside (some would say the dark side) of this increasingly fast-paced digital innovation cycle is important to think about. For example, we now create huge amounts of waste, including but not limited to electronic or e-waste. Governments have started to introduce legislation to control and reduce this (e.g., the European Union introduced an End-of-Life Vehicles Directive in 2000, which bans certain toxic materials from car production and emphasises the importance of using materials that can be recovered and reused). Furthermore, certain companies have taken action to increase the recovery and reusability of their component parts or at least make claims about the 'greenness' of their products. However, there also exists critique of such initiatives and efforts (see, for example, DesAutels and Berthon (2011)). More generally, the issue of the sustainability of life on our planet in light of the accelerated pace of innovations driven largely by digital technologies is an increasingly important challenge for organisations and their managers. This is a theme that we pick up throughout this book, although we recognise that there are no easy solutions.

An example of an innovation that does address this sustainability challenge is referred to as industrial symbiosis – this involves innovation *across* companies to reduce the net environmental impact. Industrial symbiosis (sometimes also referred to as the circular economy) involves the sharing of services, utilities and, most importantly, by-product resources among industries in order to improve the environment, while reducing costs or adding value (e.g., see Paquin and Howard-Grenville (2012)). This builds on the idea that one firm's waste is another firm's needed resource input (e.g., grain residue from the brewing process that can be used in farming to grow mushrooms rather than simply being thrown away). More generally, the mantra of 'reduce, reuse and recycle' is promoted by environmental groups to encourage consumers (and companies) to reduce the environmental consequences of increasing piles of – often toxic – waste. However, to date, little real progress has been made on this agenda and the sustainability of our increasingly 'throw-away society' remains an issue that is perpetuated by the increasing speed of innovation today.

The importance of agency in innovation

As this book has a general focus on innovation, in particular digital innovation, we return to this theme throughout. For now, the point to note is that innovation, including digital innovation, is a human activity that involves people using and creating knowledge, although, as we discuss later, this activity – or practice – also involves the *agency* of material artefacts; this last point is something that has tended to be under-emphasised in research that has focused on innovation processes. Despite this caveat, for now, we can use the definition of innovation given by Andrew Van de Ven (1986) as a starting point. He defines innovation as 'the development and implementation of new ideas by people who over time engage in transactions with others within an institutional order' (p. 590). This definition recognises the human-centric nature of innovation – it involves people interacting with each other over time to exchange ideas, learn and develop new ideas even while institutional structures can constrain (as well as at times enable) this innovation process. For example, institutional orders (the taken-for-granted ways of doing things that are not thought about because they are the only ways that are considered to be legitimate – see Scott, 1995) have arguably restrained the emergence of novel solutions to personal transport because the existing infrastructure is set up to support petrol or diesel vehicles rather than electric vehicles and vehicles that are driven by human operators rather than digitally enhanced self-driving vehicles. Existing institutional orders can create what has been described as a 'knowledge gap' – a gap between what research suggests we could do (e.g., to produce vehicles which drastically reduce vehicle emissions and autonomous vehicles that would be potentially safer) and what we actually do (e.g., continue to build and use vehicles that are less environmentally friendly and less safe than they could be) (Figure 1.1).

Figure 1.1

Developing and implementing new ideas
©Getty Images/Hero Images

In this changing context, then, where there has become a need for continuous innovation to remain competitive, there has been a gradual recognition that labour must be seen as more than simply the hands/body required to carry out physical work. Firms also need their workers' brains (their knowledge) since relying on only a small cadre (managers and scientists) for developing and introducing innovation is slow and a waste of talent. The term the 'knowledge economy' was born from this, popularised by the late Peter Drucker, initially in his book *The Age of Discontinuity* in 1969 (Drucker, 1969) but really catching on among academics in the 1990s, when the knowledge-based view of the firm emerged as a central idea in the business strategy literature (Grant, 1996; Spender, 1996). While in some ways this was a revolutionary idea – that all workers could contribute useful knowledge to add value to a firm – in other ways, the conceptualisation of knowledge work (rather than physical labour) did not change the view of people as employees that much, because knowledge has most commonly continued to be treated as a resource that can be managed – hence the increasingly widespread use of the term 'knowledge management' (Scarbrough and Swan, 2001).

> ### 🧠 Key Concepts: Innovation and the Knowledge Economy
>
> The Industrial Revolution and traditional forms of labour: land, labour and capital generated the ingredients for the Industrial Revolution together with technological innovations that allowed for mass production
>
> Scientific Management: the idea that it was possible to divide knowledge from action, so that managers would have the knowledge and decide how to do a particular job and workers would simply follow these prescriptions
>
> The service economy and the growing importance of digital innovation: the growth of services even when there are also products, has created a new type of economy with more emphasis on knowledge than physical labour
>
> The birth of Knowledge Management: seen as a product of the growing importance of continuous, often digital, innovation that requires the involvement of many

The birth of knowledge management

The term 'Knowledge Management' (KM) is used in both the popular and academic literature to refer to the general idea that organisations can generate value by improving the ways in which they create, capture/store, distribute/transfer and effectively use/apply knowledge (Grant, 1996; Choo, 1998). The focus on KM was partly a reaction to a changing business context (the growth of the knowledge economy as described above) but it was also stimulated by developments in digital technology, which led to the creation of what have been called 'Knowledge Management Systems' (KMS) (Alavi and Leidner, 2001). Particularly important has been the fact that computing power now allows us to create huge repositories of information which can be accessed from anywhere via the internet, so that people can benefit from the knowledge of others. This access is made easy because people are now digitally connected directly through computers and increasingly also a variety of mobile devices.

These digital innovations have been exploited to develop different approaches to KM. Repository KMS, for example, depend on creating repositories of knowledge that others can access across time and space, whereas network KMS rely on the ability of technology to connect knowledge workers across space so that they can share knowledge directly. These different approaches are discussed in Chapters 8 and 9, where we also consider new digital approaches to KM that rely on the emergence of social software applications and/or use (big) data that are now collected through all the sensors that increasingly track everything we do. Digital technology is therefore seen as a key element in improving knowledge processes for the extraction of business value (Tanriverdi, 2005), as we discuss more comprehensively in Chapter 8.

Knowledge exploration versus exploitation

Two basic aspects of knowledge management are typically discussed: the *exploration* and the *exploitation* of knowledge (March, 1991). Knowledge exploration refers to creating new knowledge for innovation – to produce new products, services, organisational arrangements or business models. Knowledge exploitation involves ensuring that knowledge that is potentially available within a firm and that beyond its boundaries (e.g., with the advent of crowdsourcing) it is actually accessed and used so that costly reinvention and repeating of the same mistakes are avoided. Keeping a balance between exploration and exploitation is seen to be important, so firms are advised of the importance of being 'ambidextrous' – exploiting knowledge to improve efficiency while exploring knowledge to innovate and so gain (or perhaps not lose) competitive advantage (Durcikova et al., 2011). However, for different types of firm, the balance between exploration and exploitation will be different. For instance, Hansen et al. (1999, p. 7), seeing exploration and exploitation as two distinct strategies, suggest that 'executives who try to excel at both strategies risk failing at both' and so advocate an 80:20 split – focusing 80 per cent of effort on one strategy and 20 per cent on the other. This is essentially in line with the idea that firms need to select either a cost leadership or a differentiation strategy (Porter, 1985), with cost leadership focusing on efficiency and being consistent with a repository KMS – a *codification* strategy in Hansen et al. (1999)

terms – and differentiation focusing on innovation and hence consistent with a networked KMS – that is, a *personalisation* strategy.

Firms whose business model is based on fast-paced digital innovation (e.g., Apple) or whose strategic value comes from creating unique solutions for customers (e.g., IDEO) will focus more on knowledge exploration, whereas firms that provide standard solutions to reduce costs (e.g., Walmart, McDonalds) may focus more on knowledge exploitation. However, not all subsequent research has confirmed that pursuing one or other KM strategy is always best. For example, Liu et al. (2010) concluded that pursuing an either/or approach was better when the potential from knowledge-sharing was low, but found that where the potential from knowledge-sharing was high, pursuing a combined personalisation and codification strategy was better than pursuing one or the other. Moreover, Ciborra and Andreu (2001) argue that the kinds of knowledge management approaches that are going to help within a single firm are unlikely to be successful in a network setting where firms are working closely together to create value.

In the latter context, accidental learning from spillovers is more likely to foster success, so attempts to restrict digital communication and manage knowledge transfer may end up being counterproductive even though this might prove helpful within a single firm. We expand on this point further when we discuss social software in Chapter 8. More generally, it is now also recognised that improving efficiency depends on successful innovation, as we introduced with the example of Ford's Model T. As another example, Walmart's efficiency is attributed, partly at least, to its innovation in supply chain management and its logistics developments that draw heavily on using Artificial Intelligence. This means that all firms need to be both innovative and good at knowledge exploration *and* efficient and good at knowledge exploitation. They need to be *ambidextrous* and so able to do these two apparently contradictory things simultaneously (Tushman and O'Reilly, 1996). We should note that knowledge exploration and exploitation are not easily separated in practice: knowledge exploitation relies on exploration, and knowledge exploration relies on exploitation. For example, new ideas don't appear in a vacuum, they build on what has gone before. In other words, knowledge exploration activity exploits prior knowledge. Similarly, to exploit an 'old' idea in a new context, those involved in the new context will need to adapt and explore the idea to fit that particular context.

Whether focused on exploration or exploitation, knowledge had come to be seen as a resource which needed to be 'unlocked' from employees' brains through appropriate management. And so KM emerged as a major topic of study (McInerney, 2002) and very quickly became a fashion, if not a fad (Raub and Rüling, 2001; Scarbrough and Swan, 2001). This can be demonstrated by looking at the Information Systems literature. When 'knowledge management' was used as a keyword in the Association for Information System's e-library between 1998 and 2013, the number of articles in 1998 was a mere 212, rising steadily to a peak in 2009 of 1,667 and declining thereafter.

Figure 1.2

Knowledge management in organisations
©Getty Images

Whether the fashionable term KM will survive is not certain (indeed, the above library search suggests that its popularity is waning), but the idea that knowledge (and associated data/information) is important to firms is beyond doubt. What is less certain is whether treating knowledge as a resource that can be managed like any other tangible resource is the most useful way of viewing how knowledge adds value to an organisation (Alvesson and Kärreman, 2001). Indeed, in this book, we argue that some of the problems that have been observed in research into the effectiveness of KM initiatives for digital innovation arise precisely because they are based on the assumption that knowledge can be managed and transferred, just like any other resource. We start this discussion next as we move to considering what we mean by knowledge (Figure 1.2).

Knowledge as a concept

Many discussions of knowledge include Polanyi's (1966) consideration of *tacit* knowledge as distinct from *explicit* knowledge. Explicit knowledge is that which is known and that can be codified (in words, numbers or other symbols) and is able to be communicated between those who share the same symbolic reference code. As an example, this book is written in English and so its content can be shared with an English language speaker but not with someone who cannot read or speak English. However even if you speak

English, a book or paper written in English still might not be accessible to you – for example, if it was about something very technical that you were unfamiliar with. Just try reading the technical instructions for installation of some piece of household equipment or the legal jargon in a complicated contract! So, for explicit knowledge to be shared, it depends on some overlap in knowledge between those involved – they must share the same symbolic code and have similar understandings of the phenomenon that is the subject of the communication.

Tacit knowledge relates to the idea that we know more than we can tell (Polanyi, 1966). The example that is often used to illustrate this is riding a bicycle. Once you can ride a bicycle, you 'know' how to ride, but it is not possible to tell someone else how to ride the bicycle in a way that makes them immediately competent. They have to practice themselves and 'get the feel' of riding before they can be said to 'know how to ride a bike'. A person with knowledge can share certain aspects of riding that can help. For instance, you need to push down on the pedals to turn the wheels. Unfortunately, however, a lot of what can be told does not directly help a person acquire the necessary knowledge to put it into practice. So, a knowledgeable person can explain to a novice that if you start to tip to one side, you need to shift weight the other way to restore balance, but this will not directly provide the novice with the facility to balance. Balance comes only from practice and, once learnt, this knowledge is fairly robust. Much knowledge is tacit in this sense – it is very difficult, if not impossible, to communicate to someone else in a way that directly transfers the knowledge. This means that we can see explicit and tacit knowledge not as two different types of knowledge, but as two parts of a whole: two sides of the same coin rather than two different coins. They are thus 'mutually constituted' (Tsoukas, 1996). We discuss this further below. For now, it is helpful to recognise that many KM initiatives are flawed because they rest on the assumption that tacit knowledge *can* be 'converted' into explicit knowledge (Nonaka, 1994), which is *not* something that Polanyi believed in himself. This suggests that we need to ask ourselves a broader question: *what is knowledge?*

Knowledge as the ability to discriminate

Knowledge is evident by its consequences, and this is important to consider in the context of this introductory chapter. If we think about doctors, for example, we can note that they conduct examinations of patients by using various medical devices such as a stethoscope and from this decide what is wrong with a patient and what might be the cure. A trainee may well be unable to do this, especially if the diagnosis is not straightforward. Therefore, we can say that the doctor has knowledge that the trainee does not and that the doctor is able to act knowledgeably while the trainee cannot. What distinguishes the expert and the novice is their ability to differentiate within and between the complex patterns of data and information that they draw upon in particular situations. In this sense, as Tsoukas and Vladimirou (2001, p. 979) express, knowledge is 'the individual ability to draw distinctions within a collective domain of action, based on appreciation of context or theory or both'. The ability to make distinctions and thereby draw meaning in a particular context, then, is the essence of knowledge. It is often easier to realise that we *lack* knowledge than that we *have* knowledge because when we have knowledge, the distinctions seem so obvious to us. For example, if you go to a music concert with a musician and you are not yourself a musician, you don't appreciate what you hear in the same way and so can only nod in pretence when they talk about 'the beautiful tone of the bass' or how a violinist was 'a little out of sync'. This is so because you are unable to distinguish between individual players and the rest of the orchestra. You don't hear the subtle distinctions they do. Likewise, the trainee doctor does not hear the subtle variations in the beat of the heart when listening through a stethoscope; the experienced doctor does.

Possessed and practised knowledge

Given this above definition of knowledge, the next question is: how do we gain knowledge? Philosophers have long discussed this question. While there are different views, a helpful distinction in terms of the nature of knowledge and how knowledge is acquired is provided by Cook and Brown (1999) in their distinction between the epistemology of *possession* (which treats knowledge as something individuals and groups have or own, based on prior experience but separable from that experience) and the epistemology of *practice* (which treats knowing as something people do that is context-dependent, always emerging and socially situated). Essentially, this distinction differentiates between the idea that people 'have knowledge' and that people 'act knowledgeably'. The roots of this idea that there are distinct types of knowledge can be traced back to Aristotle, who distinguished between *techne* (applied knowledge based on means–end rationality), *episteme* (theoretical knowledge) and *phronesis* (practical knowledge of how to act in the immediate situation). See Van de Ven and Johnson (2006) for a discussion of how these distinct types of knowledge create problems for the transfer of knowledge between researchers and practitioners.

The possession perspective

The possession perspective focuses on structures or routines that 'carry' knowledge and cognitions – individual mental functioning. In this perspective the mind is viewed as the carrier of knowledge or organisational rules that set out how to do something. Knowledge is, therefore, the personal property of an individual or collective knower (Spender, 1996), and mental processes are the mechanism that confers meaning – knowledge – from data and information. Checkland (1981) talks of information arising from meaning being applied to data. These mental processes are the product of experiences, perceptions and understandings, as well as theoretical learning, which create a frame of reference that allows an individual to infer particular things (i.e., to make distinctions) from some data or information. So, an individual with prior training in physics can infer meaning from the equation $E = mc^2$, which is something that someone without this possessed knowledge will be unable to do. The doctor can hear a distinction in a heartbeat using a stethoscope that tells them that the patient has a particular type of arrhythmia; a novice who does not possess this knowledge cannot. This possessed knowledge includes both tacit and explicit knowledge, as discussed above in terms of riding a bike: riding a bike involves both explicit knowledge (you must sit on the saddle, hold on to the handle bars and pedal) and tacit knowledge (how to balance). Once able to ride, both aspects of the knowledge of riding can be described as being possessed. It is important to remember in Cook and Brown's (1999) classification that both explicit and tacit knowledge are possessed knowledge while practice is needed to generate this possessed knowledge.

Possession and the practice perspective

The epistemology of practice, then, sees knowledge, or better *knowing*, as intrinsic to localised situations and practices where people perform or enact activities with a variety of others (both other people and a variety of objects) such that acting knowledgeably emerges from this practice and cannot be separated from it. Knowledge and practice are *immanent* – two sides of the same coin. In other words, knowing is not something that stands outside of practice but is constantly (re)produced as people and their tools work together, with certain consequences arising. These consequences may be more or less intentional or purposive and may demonstrate more or less 'acting knowledgeably'. The collective, or community within which practice is undertaken, is characterised by a particular set of stories, norms, representations, tools and symbols which together co-produce knowledgeable practice; it is not simply knowledge that resides in someone's brain that produces knowledgeable practice.

Being knowledgeable includes the development of a shared identity as well as shared beliefs with other members of a particular knowledge community. This shared identify and belief system are what underpin being a knowledgeable actor in any particular setting. For example, 'being' a midwife involves more than simply having knowledge about how to deliver babies: it involves a person dressing in a uniform, having the tools that will be needed ready-to-hand and being able to use them, and generally having the presence to reassure those involved that things are under control even when the birthing parents are feeling panicked (this is amply illustrated by the BBC television series *Call the Midwife*). Of course, some midwives might be classified as being more expert or 'experienced' than others, and this may reflect their possessed knowledge. However, in addition, on each occasion of 'acting as a midwife', a person may act more or less competently, not simply because of what she or he knows, but also because of how other actors interact in the emerging situation. Perhaps, for instance, the mother-to-be reacts unexpectedly to a painkiller, or some apparatus (e.g., a baby heart monitor) stops working. Situations of practice are never completely predictable, and knowledgeability includes being able to respond quickly to such occurrences (Marabelli and Newell, 2012). Nevertheless, breakdowns happen even for the most knowledgeable practitioner, and these breakdowns are useful for identifying the role and importance of the different actors involved in the practice of 'acting knowledgeably' (Sandberg and Tsoukas, 2011). Moreover, as we discuss more fully in Chapter 7, these different actors may be other human actors (expectant mothers) or non-human actors (a baby heart monitor).

Cook and Brown (1999) see these two epistemologies (possessed knowledge and knowing in practice) not as alternative or opposing views, but rather as complementary, with knowledge possessed being a tool of knowing – the doctor possesses knowledge, based on prior experience or reading about something, and this knowledge, together with the tools of his or her practice (e.g., stethoscope) and the other actors in a particular situation (e.g., the patient and the nurse), enable him/her to act more or less knowledgeably. In a different context, and without the available tools of practice that she or he uses competently to draw distinctions, the expert doctor would be little better than the novice. Possessed and practised knowledge, therefore, work together in what Cook and Brown describe as a 'generative dance'. This metaphor of a generative dance emphasises how possessed knowledge influences knowing in practice while practice changes the knowledge that is possessed – so knowledge and knowing are constantly emerging through the 'dance' of practice.

While possessed and practised knowledge work together, it is important to recognise that Cook and Brown acknowledge the central and fundamental role of practice – practice is the engine of knowledgeability. This is because an individual's (or a collective's) possessed knowledge exists only in so far as it was created using social categories derived from practice that gave sense to this knowledge (Latour, 2005). For example, the very words on this page are intelligible only because we have come to an agreement about our language, and language is the product of social action/practice. This means that possessed knowledge is always a product of past practice, just as social structures are always the product of human action (Giddens, 1984). It should be noted, however, that much of the KM and KMS literature has, more or less implicitly, adopted a 'knowledge as possession' perspective, ignoring the role of practice in generating this knowledge.

The practice view of knowledge is associated with research on particular *communities of practice*. Here, the focus is on how experts share knowledge with novices, not through a direct transfer of knowledge, but rather as novices work alongside experts so that over time they develop the skills, understanding, beliefs and, importantly, as we have seen, the sense of identity associated with the particular practice community. This is referred to as 'legitimate peripheral participation' to denote the idea that newcomers learn through gradual participation in the situation of doing/practice. The concept of communities of practice is most associated with the work of Lave and Wenger (1991) and Orr (1996).

More recently, it has been recognised that focusing on single communities may be overly restrictive because practice typically involves interactions and negotiations with people from different professions and communities (McPherson and Sauder, 2013). It is also recognised that practice is not simply a social activity but is materially mediated (as we discussed with the example relating to a doctor's use of a stethoscope and a midwife's use of a baby heart monitor). We discuss this further in Chapter 7. Ultimately, here, we want simply to emphasise that knowing is a practical accomplishment that is not restricted to the isolated cognitions of an individual, or the structural routines established in an organisation, but is rather produced in action alongside other people and objects that 'equip' (Gherardi, 2011) the environment to support the knowing.

 Key Concepts: Knowledge and Knowing

Ambidextrous: knowledge exploitation and exploration undertaken simultaneously

Tacit and explicit knowledge: tacit knowledge is that which is difficult to articulate in words, numbers or other symbols but rather is generated from experience; explicit knowledge is that which is articulated in words, numbers or other symbols which can be shared

Knowledgeability: the ability to use knowledge in practice based on both tacit and explicit knowledge

The epistemologies of possession and practice and their 'generative dance': the idea that knowledge is both a possession (of explicit and tacit knowledge) and also a practice and that it is the interaction of both that helps us understand what is going on in any given situation

Communities of practice: communities that share a common practice and can learn from each other to develop that practice at a collective level

Conclusions

We have seen in this chapter how the changing context of business organisations has put a premium on firms being more innovative, especially in relation to exploring and exploiting digital technologies. Even where the focus of competition is low cost, improving efficiency often depends on digital innovation. And innovation is fundamentally a knowledge process, involving simultaneous knowledge exploration (to identify new products, services and organisational arrangements) and knowledge exploitation (ensuring that the ideas are shared and reused across an organisation). We have also seen how knowledge can be viewed as both a possession (which includes both explicit and tacit knowledge) and a practice and that knowledgeability (acting knowledgeably in a particular situation) involves using knowledge (possessed) as a tool of knowing (in practice) that is co-produced alongside other people and things.

In the rest of the book, many of these themes are developed as we consider in more detail the relationships between knowledge, management and digital innovation; we explore these processes with case examples in other chapters. In this chapter, we end with some general discussion questions.

? DISCUSSION QUESTIONS

The following discussion questions are relevant to using this chapter in teaching exercises and discussions or for revision:

1. What do we mean by the knowledge economy and how is it different from the traditional industrial economy?

2. Why is digital innovation so important to organisations today?

3. Why is there a dark side to the more rapid product innovation cycle that is common today?

4. Why and how has the view of workers changed with the increasing importance of innovation?

5. What do you understand by knowledge and why do we need to appreciate that knowledge is both possessed and practised?

 ### Additional suggested readings

Cross, R., Gray, P., Cunningham, S., Showers, M. and Thomas, R. (2010). The collaborative organization. MIT Sloan Management Review, 52, 1, 83–90.

Davenport, T.H. and Prusak, L. (2000). Working Knowledge: How Organizations Manage What They Know. Watertown: Harvard Business Review Press.

Nonaka, I. (2007). The knowledge-creating company. HBR, July/August, 162–171.

References

Alavi, M., and Leidner, D.E. 2001. "Knowledge Management and Knowledge Management Systems: Conceptual Foundations and Research Issues," *MIS Quarterly* (25:1), pp. 107–136.

Alvesson, M., and Kärreman, D. 2001. "Odd Couple: Making Sense of the Curious Concept of Knowledge Management," *Journal of Management Studies* (38:7), pp. 995–1018.

Bolwijn, P. T., and Kumpe, T. 1990. "Manufacturing in the 1990s—Productivity, Flexibility and Innovation," *Long Range Planning* (23:4), pp. 44–57.

Brynjolfsson, E. 1993. "The Productivity Paradox of Information Technology: Review and Assessment," *Communications of the ACM* (36:12).

Brynjolfsson, E., and Hitt, L. M. 2000. "Beyond Computation: Information Technology, Organizational Transformation and Business Performance," *The Journal of Economic Perspectives* (14:4), pp. 23–48.

Checkland, P. 1981. *Systems Thinking, Systems Practice*. New York: John Wiley and Sons.

Choo, C. W. 1998. *The Knowing Organization: How Organizations Use Information to Construct Meaning, Create Knowledge, and Make Decisions*. Oxford: Oxford University Press.

Ciborra, C. U., and Andreu, R. 2001. "Sharing Knowledge across Boundaries," *Journal of Information Technology* (16:2), pp. 73–81.

Cook, S. D., and Brown, J. S. 1999. "Bridging Epistemologies: The Generative Dance between Organizational Knowledge and Organizational Knowing," *Organization Science* (10:4), pp. 381–400.

DesAutels, P., and Berthon, P. 2011. "The Pc (Polluting Computer): Forever a Tragedy of the Commons?," *The Journal of Strategic Information Systems* (20:1), pp. 113–122.

Drucker, P. 1969. *The Age of Discontinuity: Guidelines to Our Changing Society*. New Brunswick (USA) and London (UK): Transactions Publishing.

Durcikova, A., Fadel, K. J., Butler, B. S., and Galletta, D. F. 2011. "Research Note-Knowledge Exploration and Exploitation: The Impacts of Psychological Climate and Knowledge Management System Access," *Information Systems Research* (22:4), pp. 855–866.

Gherardi, S. 2011. "Organizational Learning: The Sociology of Practice," in *The Blackwell Handbook of Organizational Learning and Knowledge Management,* M. Easterby-Smith and M.A. Lyles (eds.). Oxford, Melbourne, Berlin and Malden, MA: Blackwell, pp. 43–65.

Giddens, A. 1984. *The Constitution of Society: Outline of the Theory of Structuration*. Oakland, CA: University of California Press.

Grant, R. M. 1996. "Toward a Knowledge-Based Theory of the Firm," *Strategic Management Journal* (17:S2), pp. 109–122.

Hammer, M. 1990. "Reengineering Work: Don't Automate, Obliterate," *Harvard Business Review* (68:4), pp. 104–112.

Hansen, M., Nohria, N., and Tierney, T. 1999. "What's Your Strategy for Managing Knowledge," *Harvard Business Review* (March–April), pp. 1–10.

Latour, B. 2005. *Reassembling the Social: An Introduction to Actor-Network-Theory (Clarendon Lectures in Management Studies)*. Oxford, UK: Oxford University Press.

Lave, J., and Wenger, E. 1991. *Situated Learning: Legitimate Peripheral Participation*. Cambridge: Cambridge University Press.

Liu, D., Ray, G., and Whinston, A. B. 2010. "The Interaction between Knowledge Codification and Knowledge-Sharing Networks," *Information Systems Research* (21:4), pp. 892–906.

Marabelli, M., and Newell, S. 2012. "Knowledge Risks in Organizational Networks: The Practice Perspective," *The Journal of Strategic Information Systems* (21:1), pp. 18–30.

March, J. G. 1991. "Exploration and Exploitation in Organizational Learning," *Organization Science* (2:1), pp. 71–87.

McInerney, C. 2002. "Knowledge Management and the Dynamic Nature of Knowledge," *Journal of the American Society for information Science and Technology* (53:12), pp. 1009–1018.

McPherson, C. M., and Sauder, M. 2013. "Logics in Action: Managing Institutional Complexity in a Drug Court," *Administrative Science Quarterly* (58:2), pp. 165–196.

Nonaka, I. 1994. "A Dynamic Theory of Organizational Knowledge Creation," *Organization Science* (5:1), pp. 14–37.

Oliva, R., and Kallenberg, R. 2003. "Managing the Transition from Products to Services," *International journal of Service Industry Management* (14:2), pp. 160–172.

Orr, J. E. 1996. *Talking About Machines: An Ethnography of a Modern Job*. Ithaca and London: Cornell University Press.

Paquin, R. L., and Howard-Grenville, J. 2012. "The Evolution of Facilitated Industrial Symbiosis," *Journal of Industrial Ecology* (16:1), pp. 83–93.

Polanyi, M. 1966. *The Tacit Dimension*. Chicago and London: The University of Chicago Press.

Porter, M. E., and Millar, V. E. 1985. "How Information Gives You Competitive Advantage," *Harvard Business Review* (July-August), pp. 149–152.

Raub, S., and Rüling, C.-C. 2001. "The Knowledge Management Tussle–Speech Communities and Rhetorical Strategies in the Development of Knowledge Management," *Journal of Information Technology* (16:2), pp. 113–130.

Sandberg, J., and Tsoukas, H. 2011. "Grasping the Logic of Practice: Theorizing through Practical Rationality," *Academy of Management Review* (36:2), pp. 338–360.

Scarbrough, H., and Swan, J. 2001. "Explaining the Diffusion of Knowledge Management: The Role of Fashion," British Journal of Management (12:1), pp. 3–12.

Scott, W.R. (1995). Institutions and Organizations. Thousand Oaks, CA, Sage.

Spender, J.-C. 1996. "Organizational Knowledge, Learning, and Memory: Three Concepts in Search of a Theory," Journal of Organizational Change and Management (9:1), pp. 63–78.

Tanriverdi, H. (2005). "Information Technology Relatedness, Knowledge Management Capability and Performance of Multibusiness Firms," *MIS Quarterly* (29:2), pp. 311–314.

Taylor, F. W. 1911. The Principles of Scientific Management. New York and London: Harper & Brothers Publishers.

Tsoukas, H. 1996. "The Firm as a Distributed Knowledge System: A Constructionist Approach," Strategic Management Journal (17:S2), pp. 11–25.

Tsoukas, H., and Vladimirou, E. 2001. "What Is Organizational Knowledge?," Journal of Management Studies (38:7), pp. 973–993.

Tushman, M. L., and O'Reilly, C. A. 1996. "The Ambidextrous Organizations: Managing Evolutionary and Revolutionary Change," *California Management Review* (38:4), pp. 8–30.

Van de Ven, A. H. 1986. "Central Problems in the Management of Innovation," *Management Science* (32:5), pp. 590–607.

Van de Ven, A. H., and Johnson, P. E. 2006. "Knowledge for Theory and Practice," *Academy of Management Review* (31:4), pp. 802–821.

2 THE INNOVATION PROCESS

Summary

The chapter opens with the case study of a company's (Tech Co.) innovation journey with a new digital technology system. This case is referred to throughout the chapter to illustrate the key concepts and relevant theories that are discussed. Specifically, in relation to the content, the chapter first provides an overview of innovation by 'type' (process/product/service) and by 'degree of change' (incremental/radical) and impact (disruptive or not). We then review the traditional linear view of innovation and contrast this with an interactive view that recognises that phases of innovation, such as, in simple terms, development, adoption and use of new ideas/technologies, do not occur in a straightforward sequence. They are not, in other words, stages but may be more appropriately viewed as overlapping and iterative phases. When one considers this iterative innovation process, the importance of knowledge is emphasised and the construct of *absorptive capacity* is discussed. The notion of absorptive capacity is introduced in this context given its importance in relation to the recognition, assimilation and use of new ideas (ideas that are brought into the organisation from external sources). As we have done for the linear view of innovation, we critically review how the concept of absorptive capacity is commonly viewed (as linear) and suggest that it should be viewed instead as an interactive set of interwoven phases or processes where the knowledge that an organisation 'owns' interacts with new knowledge to provide a source of digital innovation.

Learning Objectives

The learning objectives for this chapter are to:

1. Understand the characteristics of process/product/service innovation
2. Understand the differences between incremental, radical and disruptive innovation
3. Understand that innovation processes can be seen in interactive terms as well as in more traditional linear terms
4. Recognise the role of knowledge in innovation processes
5. Appreciate that, in order to create new knowledge, organisations need to interact with their environments to gain external knowledge – they need to have absorptive capacity.

Case: Enterprise system implementation at Tech Co.

Tech Co. is a worldwide company headquartered in New England in the United States. In the early 2000s, Tech Co. decided to implement its first Enterprise System (ES). An ES is a digital system based on a central database where all departments input and use standardised data, allowing these data to flow seamlessly across the organisation. In the story that follows, we present the events occurring at Tech Co. which illustrate how the first implementation was a failure and that it was only with the second implementation (or relaunch) that success was finally achieved.

The story

In 2001, Tech Co. decided to adopt an ES and selected Pluto. The decision was made by the Marketing Director and was based on what he had learnt about the benefits of an ES when he was at a marketing conference. The aim was to integrate data from different departments because the whole company was experiencing problems related to data management. It was hoped that the introduction of an ES would facilitate the common management of data; however, as we shall see, this did not happen initially. In 2001, before the ES was adopted, various data management tools were being used at Tech Co., such as Excel spreadsheets, Microsoft Word, and departmental databases. The existing systems allowed only a very individual and localised view of work. Pluto required employees 'to go put their stuff in[to] a centralised system', as a project manager involved in the implementation of Pluto described it.

The initial rollout of Pluto (2001–2003) was a failure – 'nobody was using the system' (quoting an IT manager) because it was too complicated, customised for managers rather than end users, and not properly sold to the users in terms of why they should use it. This led Tech Co. to consider abandoning the project. In 2004, there was a crisis: ROI (Return on Investment) – initially seen as a good measure of the benefits to be derived from Pluto – was not growing. The only thing that saved Pluto from being abandoned was the amount of money already invested in the project: 'Holy mackerel, we spent a lot of money!' (Marketing Director). The costs had involved the costs of Pluto itself, the hiring of an expensive consulting company, Alpha, to help with the implementation and the internal costs of people working on the Pluto project. During the initial phases of the project, they had spent a lot of time and money on customising the selected system, but in 2004, on the advice of a new consultancy company (Omega), they reverted to the 'vanilla' version, just customising small parts of the system to ensure consistent terminology or to reflect particular structures at Tech Co. Vanilla versions refer to the out-of-the-box software solution without any major customisations and relying mainly on configurations (options available in the out-of-the-box solution). Starting from early 2005, Tech Co. management felt they had achieved a better balance between customisation and configuration. In 2006, they introduced UATs (User Acceptance Tests) and in 2008 began a project to update the functionality to include the VARs (Value-Added Resellers), allowing external partners to use the ES. Finally, in order to meet the needs of different departments and facilitate integration of the system across the whole organisation, from 2007 Tech Co. focused on improving interactions across departments. Cross-department weekly working committees were established and the steering committee started having entire sessions dedicated to monitoring and evaluating the development of Pluto. All these activities were supported by an internal marketing campaign (started in 2005 with the relaunch) to sell the benefits of Pluto to users. We describe what happened in more detail next.

The failure (2001–2003)

The above narrative illustrates a number of decisions that were clearly 'wrong' for implementing an integrated system. Examples of mistakes are:

1) Very little involvement of the users

2) Emphasis on customisation, but addressed to management instead of the users

3) Lack of an internal marketing campaign

4) Focus on (short-term) financial indices such as ROI.

It is clear that Tech Co. did not initially have the necessary knowledge to undertake such a challenging ES project. However, they did ask for help – a consultancy company was hired (Alpha) that made some valuable suggestions and was very willing to 'teach' Tech Co. how to implement the system. For instance, they (Alpha) were against major customisations and made it clear that the benefits deriving from ES might be hard to measure, especially in the short term. However, Tech Co. did not seem to have been able to learn from Alpha, as they attempted a lot of customisation and focused on measuring success using short-term ROI. As one manager explains:

> We thought that when Alpha left – I think they were here for nine months, something like that – we were under the impression that we were kind of done with the implementation at that point. And as it turns out, we weren't even close.

This quote is very interesting because it suggests the incapacity of Tech Co. to absorb new external knowledge (in this case, on how to manage the implementation of a large-scale ES) even if this knowledge is clearly 'offered' by a company that has a lot of experience in these practices. In fact, Alpha was, in 2001, one of the top consultancy companies in the United States and had considerable experience in ES implementation projects.

The success (2004–2010)

The second part of the narrative shows that Tech Co. this time was able to learn from Omega (the second consultancy company, hired during the relaunch) and that they did the following:

1) Involved the users from the beginning of the relaunch

2) Started over with a 'vanilla' version of the software, which was easier to use

3) Put a lot of effort into explaining to the users why such a big change (implementing an ES) was needed (e.g., highlighting the benefits for the users)

4) Encouraged continuous cross-department discussion rather than assuming that the ES was sufficient to ensure common data practices

5) Decided to look at the opportunities offered by Pluto to make their processes more efficient in the long term. And they stopped focusing on ROI.

This second phase of the implementation is illustrative of Tech Co.'s capacity to absorb external knowledge – this time from Omega. The result was that over several years the system continued to be improved as users gained knowledge through practice and could then identify ways to

Figure 2.1

Enterprise systems in organisations

©Getty Images/Hero Images

change the system to make it even more helpful. The result was that the users were happy to use Pluto, as a manager indicates here:

> I could call them now [the users] and put them on speaker-phone, and they'd be like, 'I love this thing [Pluto]. This thing is fantastic.'

We draw on this case in this chapter as we look at processes associated with innovation, including the development, adoption and use of new ideas and things and the importance of being able to absorb external knowledge (Figure 2.1).

Introduction: Rogers's diffusion of innovation

In this chapter, we look at processes associated with innovation, including the development, adoption and use of new ideas and things. We focus in particular on digital innovation as in our opening case. However, the general processes of innovation that we discuss are broadly the same whether the innovation involves digital technologies or not. We start by providing an overview of what the new ideas and things are and point to the difference between process, product and service innovation and between incremental, radical and disruptive innovation (in so doing, we point to the blurred boundaries between these 'types' of innovation). We go on to explain and explore Everett Rogers's (1983) model of (the diffusion of) innovation. Rogers explains and potentially predicts how, for what reason and to which extent new things and ideas (involving products, processes and/or services) spread across individuals and organisations. Rogers's *Diffusion of Innovation* was first published in the early 1960s and has been revised over a period of 40 years. Rogers's work has several merits. For instance, his model suggests that 'central organisations' can promote and so facilitate more rapid diffusion of innovations to communities of potential adopters, and this approach represents a good predictor of the rate of adoption of new ideas and things over time and space. However, in this chapter, we point to some shortcomings associated with the model's linear assumptions that make the innovation diffusion process appear to be largely 'predictable', when in fact this is often not the case.

To Pre-empt, there is a clear gap between Rogers's traditional model of the diffusion of innovation and what often happens in organisational settings – the Tech Co. case above, for example, illustrates that innovation is not always a smooth process. Therefore, in this chapter, we contrast Rogers's view with a more practice-oriented perspective that accounts for the unpredictable nature of how innovations spread across a particular community and within individual firms. In the final part of the chapter, we introduce the concept of *absorptive capacity*, which highlights that innovation derives from the appropriation of 'new' external knowledge but we then explore how this appropriation is not always straightforward. We discuss absorptive capacity because the diffusion of innovation is based on how far the recipients of a new idea or thing – the adopters – have sufficient knowledge to understand what they are adopting and how it can be utilised.

Next, we provide an overview of innovation, identifying product, process and service innovation and highlighting that innovation can be more or less incremental as opposed to radical and sometimes can be disruptive.

Innovation: An overview

Innovation is a complex process involving the development of a new idea and/or 'thing' and its adoption and use among a community of potential users. Innovation and the idea and/or thing may be new to individuals, organisations or indeed to the world. Furthermore, innovation might vary between something that is new to some, and not to others, or that is completely revolutionary. For example, we treat the Tech Co. case as a digital innovation case because while many organisations before Tech Co. had introduced an ES, the ES was new to Tech Co.

We next consider different types of innovation, pointing to the differences between process and product innovation and then between incremental, radical and disruptive innovation. We also consider service innovation as a unique and increasingly important type of innovation.

Product and process innovation

When we think about innovation, we often consider the development and introduction of a product that is new to the market – Apple's first iPhone or Ford's Model T as the first affordable car. While these are certainly innovative product examples, many innovations do not involve the development of a brand-new type of product; innovation can also involve small changes to an existing product, as with the various new versions of the iPhone that Apple has released over the years.

Moreover, innovation can also be about developing and introducing a new way of producing products – this is referred to as process innovation. The Tech Co. case is an example of a process innovation – they were not trying to introduce any new products to the market; rather, they were trying to improve the way that they managed their internal processes so that their products and services could be more effectively produced and delivered. Thus, process innovation involves the introduction of new elements into an organisation's operations, such as task specifications, work and information flow mechanisms, and the equipment used to produce a product or render a service (Swanson, 1994).

Business Process Reengineering (BPR) is an example of process innovation. BPR involves rethinking organisational processes to improve efficiency and effectiveness, typically using digital technologies to support the newly designed processes. The idea is to ignore the way things were previously done, starting instead with a blank sheet of paper. Hammer (1990), for example, proposes 'obliteration' when redesigning business processes on the back of technology. For instance, FedEx developed and introduced a digital tracking system so that customers could see where their goods were in the delivery process. This is a good example of BPR as the company was focusing on the objective of the process – pleasing customers by allowing them to know what was going on with their deliveries – and this led them to a complete overhaul of their existing processes rather than simply reviewing and improving them (Figure 2.2).

Another example of process innovation relates to using the Six Sigma system to review and improve a business process – for instance, by using the DMAIC (Define, Measure, Analyse, Improve and Control) methodology. In contrast with BPR, Six Sigma is an example of a Business Process Improvement (BPI). Specifically, Six Sigma is a methodology involving a set of techniques aimed at reducing process variability, so that each time a certain process is executed, it will deliver the exact same output. For instance, the production of computer components needs to be very accurate since a hardware part that is made slightly larger, or smaller, than the required standard cannot be installed on a (PC/laptop) motherboard. Six Sigma, therefore, leads to improvements of existing production processes. While initially Six Sigma was used in manufacturing to reduce product defects by identifying the root cause of those defects, this methodology is now widely adopted across sectors, including service sector organisations such as healthcare.

Figure 2.2

Tracking transportation of goods
©Getty Images/EyeEm

While product and process innovations have different aims, it is also important to recognise that they are generally interwoven, meaning that product innovation usually involves new processes and vice versa. So, the Model T car was possible only because of innovations involving the introduction of assembly line technology (i.e., process innovation) that allowed the car to be made with lower production costs than previous systems of production.

Incremental, radical and disruptive innovation

It is clear that both process and product innovation might have a positive impact on organisational performance (e.g., sales and revenues in a for-profit company), which is why organisations want to innovate. However, the examples in this chapter illustrate that the extent of the innovation can be more radical (the first iPhone and BPR) or more incremental (new iPhone versions and BPI). In other words, some innovations build on what has gone before whereas others involve more radical departures from existing products and services. For instance, the car, as an innovative product when it was first introduced to the market, would be described as a radical innovation because it led to major changes, for example in terms of the demise of horse-drawn carriages and later railway lines, the increase in the number of roads and generally people's ability to move around their environment for pleasure or work. Later, 'new' car models built on this but were incremental in terms of their improvements – making the car more comfortable, faster, safer, able to take more people and so on.

However even within the category 'car', there can be radical innovations, as with the current focus on designing autonomous vehicles that do not require a human driver. A good earlier example of a radical 'car' innovation was the single-seater C5 battery vehicle that was designed by Clive Sinclair in 1985. This actually turned out to be a commercial disaster. As such, we need to remember that pursuing radical innovation is risky because we do not know how people will react to it. And it is not only risky for the company producing the radically different product. It can also be risky for early adopters since if the new product does not 'take-off' they might have invested in something that quickly becomes obsolete. For instance, think of the mobile companies that in 2007–2008 decided to commercialise video cell phones such as AT&T. This digital innovation was not ready for diffusion because of limited internet bandwidth, which meant that video reception was very poor. This was a product failure for AT&T but it was also bad for early adopters who invested in the video-phone but could not really use it given that few others had the video-phone.

We should also note that incremental innovations can be competitively very important even though they 'merely' build on and aim to improve previous products or services. For instance, Apple continues to upgrade the iPhone, first by making it more 'sexy' (Arruda-Filho et al., 2010), following Steve Jobs's philosophy that 'smaller is better', and then in 2014, by commercialising a larger model: the iPhone 6 plus, to compete with a similar device produced by Samsung – the Galaxy 6S.

Radical innovations can also be *disruptive* if successful (Christensen, 1997). Disruptive innovations are named as such because they disrupt the competitive equilibrium, threatening companies that produce products or offer services that are made redundant by the radically new product/service. We saw this in the car example, which was disruptive for those who produced horse-drawn carriages, for horses (and their keepers) and for the railway industry – think of all the lines that were closed because people preferred to drive themselves rather than use a train (ironically, now this trend is reversing given that there are too many cars on the road so that driving can be very slow, especially in rush-hour). Another (historical) example of a disruptive innovation is the introduction of optical supports for data storage (compact disks, or CDs), which saw access to data reconsidered from scratch (Figure 2.3). This represented a disruptive digital market innovation as it put floppy disk manufacturers out of business. In addition, it was disruptive because it affected the computer market as computer vendors started to reconsider their component design, and programmers could write larger portions of code for operating systems, for example. More recently, Airbnb has been described as a disruptive innovation because of the threat it represents to the hotel industry.

In sum, if we look at the *output*, we can identify 'types' of innovation, namely process and product innovation. If we look at *degree of change* – and

Figure 2.3

Modern laser disk (CD) reader
©*GETTY*

Table 2.1 Types of innovation

	Process innovation	**Product innovation**
Radical innovation	Business Process Reengineering	iPhone
Incremental innovation	Six Sigma	iPhone generations
Disruptive innovation	assembly line	compact disks

the risk involved, for organisations but also for consumers and society – we can distinguish between incremental and radical innovation. If we consider the *impact* on the existing industry structure, we can identify *disruptive* innovations. Incremental, radical and disruptive innovations apply to product and process innovation. Table 2.1 summarises the above discussion.

Service innovation

We mentioned in Chapter 1 the idea of service innovation, but we need to look more specifically at what this means. Service innovation is an increasingly important type of innovation because many new digital products come with a range of services that not only provide added value to customers but are important sources of revenue for the companies introducing them (e.g., Apple's iTunes, which provides iPhone users with a range of services, is integral to the success of the iPhone – the product). More generally, the service sector now accounts for a significant proportion of economic activity so that service innovation, typically incorporating digital technologies, is increasingly important.

Service innovations can also be described as being more or less radical/incremental. Dropbox is an example of a digital service innovation. Dropbox allows the remote storage of files so that people can access and share files wherever they are from whatever device they are on, as long as they have access to the internet; but they can also sync their device with the remote storage so that they can work on files while not connected, with the files automatically updating the remote storage when again connected to the internet. From the firm perspective, this example may be considered an incremental innovation; file sharing applications have been around for decades and did not require development of new technologies. From the customer perspective, however, the digital innovation is more radical; ubiquitous access has radically changed user practices, allowing people to work anywhere and at any time and to easily collaborate and share and work together on files.

Conversely, some digital service innovations may be radical (and risky) for organisations but might be perceived as incremental innovations by their customers. Think of a cable company that implements a Customer Relationship Management (CRM) system for a call centre that manages technical support. Whereas customers will (hopefully) perceive an improvement in the company's responsiveness to their queries, company employees will experience this as a more radical change given the major changes in their daily routines that will be involved, as in the Tech Co. case. Table 2.2 provides a summary.

 Key Concepts: Types of Innovation

(Digital) product innovation: development, adoption and use of a new product (e.g., a digitally enhanced car)

(Digital) service innovation: development, adoption and use of a new service (e.g., an online car loan service)

(Digital) process innovation: development, adoption and use of a new way to produce products or services (e.g., a robotically controlled assembly line to produce a car)

(Digital) incremental innovation: improvements to existing products, processes and services (e.g., a new digital entertainment system for a car)

(continued)

(Digital) radical innovation: brand-new products, processes and services (e.g., autonomous (self-driving) vehicles)

(Digital) disruptive innovation: brand-new products, processes and/or services that threaten incumbent producers of products, processes and services (e.g., idea of a digitally controlled vacuum tube to transport people and goods)

Table 2.2 Firm/user perspectives of innovation

Examples of innovation and impact for firms and users	Firm perspective	User perspective
Dropbox (example of a digital service innovation)	Technological innovation in server farms to manage customer storage that involves increased capacity, more security, stable internet connection, business continuity and backup plans.	Revolutionary way to work because of ubiquitous access to all documents from anywhere.
iPhone 5 and 6 (example of a digital product innovation)	Radical changes in the technology: new IOS; new processor (A6); touch sensor rebuilt and no longer incorporated into the display, making the phone thinner; motherboard redesigned (charger socket changes position as a consequence of this design).	Few aesthetic improvements; the phone has more RAM so it is faster.
CRM or ES introduction (example of digital process innovation)	Employees need to learn a brand-new system; this can be challenging, especially if the organisation previously used a non-centralised system.	Incremental improvement. For example, the customer does not need to spell out first and last name because the CRM recognises the source number and associates it with a specific customer – saving the customer time.

While in this section we have provided a high-level overview of innovation, we now focus on two main approaches to innovation. The first is the 'traditional' approach, which portrays innovation as a linear process (Rogers's theory of innovation). The second suggests that innovation is better viewed as an iterative, interactive process. First, we describe Rogers's model.

Rogers's model of the diffusion of innovation

Rogers (1983) identified a number of key concepts that help explain how new ideas or things spread or diffuse. Obviously, diffusion of new ideas or things is important because if the idea/thing does not spread, it can't be described as a successful innovation (as with the C5 battery car, previously mentioned). The concepts in his model can be seen as belonging to two distinct categories: *resources* (antecedents that play a role in the diffusion of innovation) and *processes* (describing how diffusion occurs).

Antecedents of the diffusion of innovation

In terms of resources, Rogers first considers *the innovation* itself – the new product or service. According to a broad definition of innovation, the new idea or thing should always be about something involving a positive change, or 'the application of better solutions that meet new requirements, unarticulated needs, or existing market needs' (http://en.wikipedia.org/wiki/Innovation). Unfortunately, sometimes new ideas/things can have less positive applications/consequences. For example, some might argue that the idea to collide two or more atomic nuclei at a very high speed in order to form a new type of atomic nucleus to produce energy was not always employed in ways that benefitted society. This example illustrates how sometimes there are unintended consequences of the creation of new ideas and things that can have negative as well as positive outcomes. Indeed, our previous example of the car as a disruptive innovation makes it clear that there are unintended negative consequences from this new (at the time) product, especially in relation to pollution but also now in terms of congestion that threatens free movement – originally the main benefit. This example also illustrates that the consequences of innovation may take a long time to manifest.

Despite this caveat, for Rogers the new idea/thing would diffuse depending on whether it provided clearly superior benefits (relative advantage) as compared to existing ideas/things. For instance, he considered the diffusion of new varieties of agricultural crops (e.g., new varieties that could better withstand drought conditions) and concluded that the greater the benefits of the new variety (i.e., they produced a higher crop yield), the quicker and further would the new variety diffuse (that is, be adopted by more farmers). As another example, we can think of

IBM's failure to successfully commercialise the operating system OS2, the first system that used icons and a mouse for PCs. Even though OS2 was equipped with a graphical user interface (GUI) – a novelty for the early 1980s – the system itself was complicated to use (with lengthy manuals and the need for considerable technical training). Around the same time, Microsoft 3.1, based on the same philosophy (GUI and mouse), was seen to be 'fool proof' and succeeded. That is, it was very simple to use and anyone with little or no prior knowledge of computers could easily move files into folders, use a control panel to manage the basic configurations of the operating system and so on. In this example, then, the Microsoft system was clearly superior in terms of the benefits to users (even if not technically superior) and this accounted for its greater diffusion. The example also illustrates that how much benefit or advantage a new product or service provides depends on the adopters; if potential adopters don't have the knowledge and skills needed to use the new idea or thing, they will not see much benefit and so diffusion will be restricted. Rogers includes other elements that contribute to the determination of relative advantage, such as how different it is to existing solutions and how easy it is to trial and so get personal understanding of the added benefits (or not). Interestingly, an ES, as adopted in Tech Co., is not easy to trial and is quite different from existing processes and so one might think that ES would diffuse slowly. However, the proposed benefits of an ES are clearly articulated by ES suppliers and consultants even though in practice these benefits are not always easy to achieve, as the case illustrates. This illustrates the role of the supply side in promoting the benefits of a particular new product, process or service – even when these benefits may be elusive to exploit in practice.

Rogers's second resource concerns *adopters*. Adopters play a key role in translating an idea into a valuable innovation; we will see how the adoption process unfolds when we examine the process elements. Adopters can be customers (e.g., individuals who buy a smartphone, the innovation), organisations (e.g., Tech Co. described earlier) and even countries (e.g., that decide to build tracks and trains to improve transport).

The third resource concerns *communication channels*. In order for an innovation to diffuse, effective communication channels need to be in place and support the transfer of knowledge (i.e., about the new idea or thing and its applications and advantages). These communication channels rely initially on mass communication as a means of broadcasting so that many become aware of the new idea or thing. Later, personal communication networks become more important since this type of channel is often needed to persuade a specific person or organisation to actually adopt. Fashion-setting organisations (Abrahamson and Fairchild, 1999) play a particularly important role in communicating new ideas and things (e.g., consultancy companies are often active at practitioner conferences presenting the latest 'new best' idea). The Tech Co. case illustrates this.

The original decision was made by the marketing director following a conference that he had attended where speakers had extolled the benefits of an ES, thus illustrating the power of conference presentations.

In the foregoing, we have reviewed the resources that, according to Rogers, are required for a new idea or thing to be widely diffused. Next, we look at the processes of innovation that account for how this diffusion happens over time.

Processes and the diffusion of innovation

According to Rogers, innovation diffusion happens in a fairly predictable way as represented by the S-shared diffusion curve. This suggests that new ideas/things first spread rather slowly, then gather pace (if they are successful) and diffuse widely, before levelling off (and potentially declining thereafter). This process is affected by the *social system* and specifically by different 'types' of adopters. He defines the following categories of adopters:

1) *Innovators* are customers (who may be individuals or organisations) who have financial capacity, influence over their social system and tight links with other innovators and are risk-takers.

2) *Early adopters* are customers with financial capacity, like innovators, but who exhibit more opinion leadership and discretion than innovators. They are a key link in the chain of innovation as they provide candid feedback to the creators of the innovation that will help improve a product/service to make it more marketable.

3) The *early majority* have above-average social status, like early adopters, but they adopt an innovation after a period of time. While early adopters can be socially seen as 'cool', the early majority includes a variety of customers who are not necessarily risk-takers and rarely hold a social position as opinion leader. Tech Co. could best be described as falling into this category as they were fairly early adopters of an ES compared to some of their competitors but certainly not among the first.

4) The *late majority* are sceptical about innovations, have below-average status and limited financial resources and exhibit very little opinion leadership. They adopt an innovation once a wide number of other customers have purchased, utilised and fully exploited the innovation – they become customers when an innovation is 'risk-free'.

5) *Laggards* have a limited social network – their social interactions generally span from family to a 'circle' of close friends – have very little financial capacity and no opinion leadership and are generally anchored to traditions and therefore show a high degree of resistance to change.

Of course, this S-shaped curve may not always be completed. Therefore, while early adopters might gain a competitive advantage by being the first to adopt, they also risk suffering losses if, for example, adopting a certain innovation requires fixed-cost investments but then the innovation is not well received by others and so there is no subsequent diffusion. We saw this earlier with the example of the video-phone. At other times, the S-shaped curve may take a long time to emerge. As another example, Clive Sinclair's son, Crispin, is now developing a new version of the personal electric vehicle. Perhaps this time there will be more perceived benefits for users so that this product will now be more successful, but of course, only time will tell. Rogers ultimately provides percentages of customers who fall into each category: innovators (2.5 per cent), early adopters (13.5 per cent), early majority (34 per cent) as well as the late majority (34 per cent) and laggards (16 per cent).

While diffusion of new ideas and things, then, generally occurs over time (accounting for a predictable diffusion curve), it is worth noting that, today, digital innovations that incorporate 'programmability, addressability, sense-ability, communicability, memoriseability, traceability, and associability' (Yoo, 2010, p. 226) can experience much faster diffusion. Digital innovation creates positive network externalities that speed up the creation and availability of devices, networks services and contents (Hanseth and Lyytinen, 2010). An example is WeCash, an online credit rating and loan company which managed to grow from 0 to 600,000 users in just eight months (Huang et al., 2017). Moreover, today, social media can speed up diffusion processes. For instance, early adopters can post their opinion about the latest smartphone, celebrating or criticising new functionalities. In this way, the message can be spread quickly and people can become opinion leaders in their social media network, either encouraging faster diffusion or reducing diffusion. We will return to this topic on the relationship between companies and consumers through social media in Chapter 8. For now, it is enough to understand the relevance of communication channels for the diffusion of innovation and to be aware that internet and social media are changing communication dynamics within and between organisations and consumers, subsequently altering the speed with which digital innovation diffusion happens.

Reflections on Rogers's model of innovation

Rogers's model of the diffusion of innovation provides a number of very interesting insights. For instance, it is helpful to understand the trade-off between being an early adopter (an organisation being a first mover, and potentially with an initial competitive advantage but also with high risks) and a late adopter (an organisation acting as a follower, with the associated risks and benefits). Furthermore, Rogers's theory accounts for a quite comprehensive set of 'variables' affecting innovation that include characteristics of the new idea/thing itself (relative advantage), the market/demand (the categories of adopters) and communication channels (mass and personal).

Key Concepts: Rogers's Theory of Innovation

Resources:

The innovation – the new product, process or service that is being introduced and the relative advantage of this over existing products, processes and services

Adopters – those individuals or organisations that decide to adopt the innovation

Communications channels – the networks that link producers of the innovation with adopters and include central organisations for mass communication and personal networks, including early adopters, for influencing individual adopters

Process:

Social systems of adopters – includes innovators, early adopters, the early majority, the late majority and laggards who adopt the innovation either late or not at all

Time – the S-shaped diffusion curve where at first adoption is slow, then picks up as more adopter's influence others and then declines as the market is saturated

As we pointed out in the introduction to this chapter, Rogers's theory portrays innovation as a linear process where things happen in sequence. For instance, once an idea or thing has been developed, it is considered to be static – so a specific product, process or service is diffused. In practice, products, processes and services change following adoption. Moreover, the linear view assumes that adoption is equated with use – so it is assumed that those who have adopted will follow through and use and so exploit the potential of the innovation; after all, they have adopted because of the relative benefits of the new idea or thing. Yet, as we have seen in the Tech Co. case, adoption does not always equate to use – in Tech Co. they adopted Pluto, but it was not used widely, at least initially. Next, we illustrate some of the challenges associated with innovation diffusion and exploitation that suggest a more interactive (and less linear) view of how innovation unfolds.

An interactive view of innovation

An interactive view of innovation suggests that innovation phases (development, adoption and use) do not necessarily occur one after the other and instead illustrates that, very often, back-and-forth interactions are needed. For example, software such as Windows 95 or the first IOS for iPhone has undertaken major modifications to become 'stable', and these modifications were possible only because of their initial use by members of the public, so

Linear view:

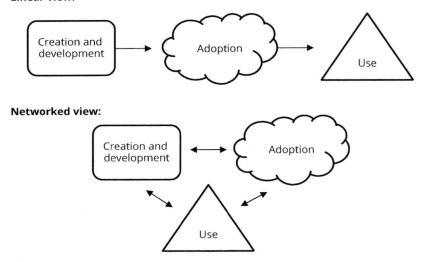

Networked view:

Figure 2.4

Linear and interactive views of innovation

use fed-back into further development, which then affected further adoption. Figure 2.4 highlights the difference between the linear and the interactive view.

The interactive view suggests that development, adoption and use do not follow one from the other but instead that there are back-and-forth movements between these phases. An example from the healthcare industry is illustrative of this interactivity. According to the linear model of innovation, the identification of a new therapy that can effectively and efficiently treat a disease would diffuse smoothly (because it has clear benefits that are easily communicated) and once it was diffused (i.e., adopted) the assumption would be that it would be straightforwardly used. So, once a new therapy is developed (that is, the new therapy has gone through the extensive regulatory process of development) and is proven to be successful and cost-effective, health practitioners would adopt and then use the new therapy. However, this is not always the case in practice. For example, brachytherapy was a proven new treatment for prostate cancer. Brachytherapy involves inserting radioactive seeds into the cancerous tumour, killing the tumour from the inside. The new treatment was slow to diffuse and was not always used even when it was adopted in a hospital. How do we explain this? At the time that brachytherapy was introduced, the dominant therapy involved surgeons cutting the tumour out. With brachytherapy, the role of surgeons in the treatment of prostate cancer becomes less dominant while radiologists are given a greater role (a surgeon will still be involved as she or he has to make the cut to insert the seeds, but the radiologist must decide and manage the dose). In this situation, radiologists were keen to adopt while surgeons resisted and so thwarted use. In this case, then, adoption by one group of physicians failed to translate into use, because of resistance from another group of physicians. This illustrates how the social system can inhibit the straightforward progress from adoption to use even when a new product has demonstrable benefits.

As already mentioned, the Tech Co. case is also a good example illustrating the interactivity between development, adoption and use. With this type of ES technology, there are some development decisions that follow adoption because the technology cannot simply be 'plugged in' but must be configured and/or customised to suit the particular context of adoption. Moreover, once these development decisions have been made and the system (in this case, Pluto) 'goes live', use does not always follow adoption. Therefore, in the Tech Co. case even though the system was ready to be used, as it had been installed on the employees' computers, it was not immediately used and exploited to the full. Indeed, in this case, the lack of use nearly led the company to abandon the system. Unfortunately, this case is not unusual since it is well known that enterprise system failures (when a firm is not able to initially exploit an adopted system) are frequent. In the Tech Co. case, they did not abandon, but rather went for a relaunch, yet the case nevertheless illustrates that some development decisions follow adoption, and these development decisions in turn can influence use, meaning that use does not always follow adoption in a straightforward manner. Furthermore, use itself can feedback into development, with modifications to the system happening after it is initially rolled out, based on users learning what works and what can be improved.

The linear view of innovation, therefore, is insufficient to explain how innovation processes unfold because it fails to acknowledge the often back-and-forth movements between the phases. The resistances in both the Tech Co. case and the brachytherapy example are illustrative of the challenges inherent in many innovation contexts. One way of understanding these challenges is to consider them from a knowledge perspective. An innovation will necessarily involve having to understand some new knowledge while challenging some existing knowledge: so the surgeons had to understand something about the advantages of radioactive seeds while also admitting that their own knowledge and skills were not sufficient to best treat prostate cancer. Similarly, the Tech Co. users had to understand how and why a centralised database may be helpful while foregoing their knowledge about and use of their own, often self-developed, local systems. These examples highlight the relationship between innovation and the knowledge that is needed to successfully appropriate a new idea or thing. To this end, we next introduce a knowledge-based framework that shows the challenges associated with the diffusion of innovation; this highlights the relevance of knowledge and knowing.

Absorptive capacity

The Role of Knowledge in Fostering Innovation

Knowledge and learning are at the heart of the development, adoption and use of new products, processes and services (i.e., the innovation process). Absorptive capacity is one framework that emphasises this point.

Innovation and absorptive capacity

Absorptive capacity stresses the relevance of prior knowledge as a necessary condition to absorb (new) external knowledge relevant for innovation. The construct of absorptive capacity distinguishes between what we already know (prior knowledge) and what we aim to learn (external knowledge, which might represent a source of innovation). Firms can develop new products, processes or services by combining existing organisational knowledge with new external knowledge (Tsai, 2001). However, while external knowledge has been shown to be a source of innovation, capturing external knowledge should not be viewed as a taken-for-granted process: it depends on the absorptive capacity of the receiving entity.

The original absorptive capacity concept was developed by Cohen and Levinthal (1990). Absorptive capacity was seen as an organisational resource that is needed to identify relevant (external) knowledge, then to understand and assimilate this knowledge, and apply it for commercial use through R&D. The most important element of absorptive capacity is that it highlights the relevance of prior knowledge – a certain 'stock' of internal knowledge – as a necessary precondition to being able to recognise and then assimilate and apply external knowledge. This prior stock of internal knowledge needs to be somewhat related to the knowledge that is to be absorbed; otherwise, the recipient organisation is not able to understand its relevance nor assimilate and use it effectively. This view of absorptive capacity is illustrated in Figure 2.5.

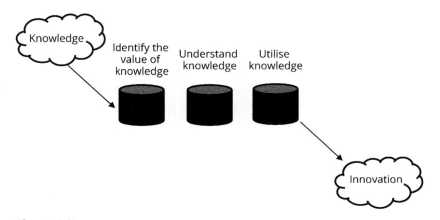

Figure 2.5

Key steps in absorptive capacity

The absorptive capacity phases figure illustrates the linearity of the model. External knowledge needs to be *recognised* as relevant, then *assimilated*; it needs to be shared across business units and 'understood' at the organisational level. This knowledge is then combined with existing organisational knowledge and *applied* to create innovation. For instance, Lane and Lubatkin (1998) undertook a study of the creation of new drugs and other biotechnical innovations in pharmaceutical and biotechnology R&D alliances and measured absorptive capacity in terms of the companies' ability to identify relevant knowledge and assimilate and apply it. They found that greater levels of absorptive capacity were reflected in higher levels of innovation.

Process view of absorptive capacity

This view of absorptive capacity was substantially revised in the 2000s. The most notable reconceptualisation was developed by Zahra and George (2002), who suggested that absorptive capacity should be seen not only as a resource – the capacity to identify, assimilate and apply external knowledge – but also as a dynamic learning capability that needs to be developed over time. This capability is broken down into two stages: knowledge *exploration* (including a phase of recognition and a phase of assimilation of new external knowledge) and knowledge *exploitation* (including transformation and application of external knowledge) – as discussed in Chapter 1.

This model has two main advantages. First, it sees absorptive capacity as a set of dynamic processes rather than phases. Second, it incorporates the concept of learning, which, on the one hand, highlights the cumulative and path-dependent characteristics of absorptive capacity and, on the other, provides the opportunity to use the construct to investigate such learning processes that do not involve just 'discovering' or creating new things through scientific research, as was the early focus of research on this topic. Therefore, a critique that Zahra and George level against Cohen and Levinthal's original model is that their three-phase model was applicable almost exclusively to R&D contexts. Zahra and George's model, instead, can be applied to innovations that are 'appropriated' by an organisation and have their main applications within the organisation. As such, there is a shift from a focus on *outputs* (a scientific finding, through R&D) to *learning* (as with Tech Co. and their ability to learn to use Pluto). This model would suggest that an organisation that aims to 'learn' to use an ES should have some prior relevant knowledge about such centralised systems that would allow it to identify and assimilate (i.e., explore) relevant external knowledge (e.g., with the help of a consultant company with relevant expertise), eventually being able to transform and apply (exploit) this knowledge in its new practices. Our interpretation of the process of absorptive capacity is illustrated in Figure 2.6.

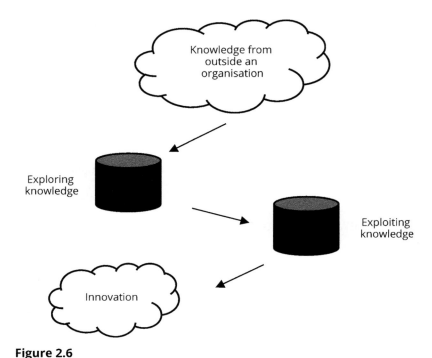

Figure 2.6

Example process of absorptive capacity

Interactive view of absorptive capacity

Figure 2.6 clarifies how organisations can bring in new ideas (i.e., use external knowledge) through learning processes, thereby going beyond the limited application of absorptive capacity – to R&D units – originally considered by Cohen and Levinthal. However, the linearity of the model remains: the model suggests that, first, knowledge exploration occurs, and then, once exploration is over, the organisation can focus on exploiting previously recognised and assimilated knowledge. As we did in contrasting Rogers's model with the interactive model, here we would like to highlight the potential of seeing absorptive capacity as an *interactive* process, where knowledge exploration does not have to be 'over' for the organisation to move on to a next step – knowledge exploitation.

The interactive nature of knowledge exploration and exploitation in the absorptive capacity construct can be understood if we think of the relationship between knowledge (possessed) and knowing (practice), as covered in Chapter 1. Some prior (relevant) knowledge – the possessed knowledge in Cook and Brown's (1999) terms – is the tool that organisations need in order to recognise the value of exploring external knowledge. When this knowledge is brought into the organisation, it needs to be assimilated and shared across business units and practised through social interactions, with knowledge acting as a tool of *knowing*, translated (or transformed, in Zahra and George's terms) into *organisational learning* (see also Sun and Anderson, 2010). However, we need to remember that this learning does not occur all at once. Instead, consistent with Cook and Brown's 'generative dance' of knowledge and knowing (again, see Chapter 1), there is a continuous back-and-forth between knowing, which originates through practice, and knowledge, which becomes possessed.

Hopefully, you are beginning to realise that it is unrealistic to think that an organisation can 'learn', for example, how to reap the benefits of an ES in two phases (exploration followed by exploitation) without making mistakes, revising previous practices and refining learning. Ultimately, the Tech Co. case shows how practices associated with the adoption and use of an enterprise system can be seen as a never-ending learning process (Markus and Tanis, 2000): the more an organisation exploits its knowledge to use a new technology, the more possessed knowledge is created and the more new knowledge can be explored (e.g., how to use the system in a unique manner that produces more benefits than those originally envisaged). Figure 2.7 illustrates our interpretation of an interactive model of absorptive capacity.

Figure 2.7 can be seen as a means of demonstrating that knowledge exploration and exploitation interact within an organisation. This is interesting because it underlines the point that organisations need to be able to

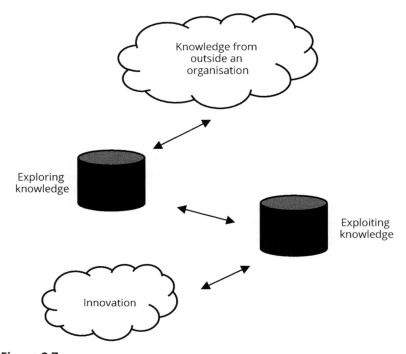

Figure 2.7

An interactive view of absorptive capacity

undertake both exploratory and exploitative activities at the same time. The idea that knowledge exploration and exploitation are overlapping processes is supported by the literature on ambidextrous organisations, as noted in Chapter 1. The idea of ambidexterity, together with our interactive view of absorptive capacity, assumes that exploration and exploitation can support each other, feeding from and to each other iteratively over time. However, there are also tensions between these knowledge processes that will be discussed in later chapters – for instance, in Chapter 4 on strategy, in which we look at ambidexterity as a *strategising* practice.

🐑 Key Concepts: Interactive View of Innovation and Absorptive Capacity (AC)

Linear model of AC:
Knowledge exploration, *then* exploitation is seen as a sequential learning process

Interactive mode of AC:
Knowledge exploration *and* exploitation is seen as an interactive learning process

We conclude this chapter by illustrating an additional issue that can influence absorptive capacity and so innovation processes – namely power.

Power and knowledge absorption

Although power is a complex construct that we cannot fully explore in this book, we need to remember that power is a key concept in the knowledge management literature, and we will see how it relates to knowledge in later chapters. For now, it is important to know that knowledge and power are very tightly related since power relationships can condition how knowledge is managed. The brachytherapy example earlier in this chapter illustrates how power dynamics can influence the innovation process. Think, for instance, of the power of a top manager/director in forming teams, thereby conditioning what knowledge will be shared among team members given their more or less common background and particular skill sets. Or think of the relevance of power when it is associated with control, so that if employees feel that they are controlled they are less likely to take initiative or propose new ideas (Podsakoff et al., 1982). On the contrary, it has been shown that empowering individuals – and so using power in a positive way – promotes a sense of identity with an organisation or project (Hasan and Subhani, 2012). Pansardi (2012) points to the difference between 'power-over' – power that derives from the organisational *hierarchy* – and 'power-to' – or *empowerment*. This distinction is important because absorptive capacity is strongly influenced by how power is exercised. In Tech Co., we see the relevance of this in the sense that during the initial launch of Pluto there was no real user involvement and users were simply told that they should now use the new system (power-over was the dominant approach and was unsuccessful). During the relaunch, on the other hand, users were involved and customisations were made with their needs in mind (rather than the managers' needs which dominated the initial launch) and they then began to use Pluto (and indeed found it to be valuable).

Conclusions

In this chapter, we have discussed different types of innovation but then focused on innovation as a process, contrasting two main perspectives: the *linear* view and the *interactive* view. It is important to be familiar with the linear view because it offers a very clear model to predict how diffusion of innovation occurs, through phases. The diffusion of innovation (even if the idea or thing that is diffusing is brilliant and clearly superior to existing solutions) does not occur overnight, but instead is often a long-term process. However, we also considered how, nowadays, digital technology and especially social media are changing the pace of innovation since social networks are communication channels where ideas and things are discussed, reviewed and praised or criticised by the public very quickly. While the linear model is therefore helpful, we also pointed to problems associated with this view, suggesting that viewing innovation processes as *interactive*, rather than straightforward and taking place step-by-step, better reflects what actually happens when an idea or thing is first developed, then refined, implemented and ultimately exploited by large numbers of individuals or organisations. Finally, we introduced the construct of *absorptive capacity* that illustrates the relevance for organisations of developing capabilities that support knowledge exploration and exploitation in order to generate innovations. We illustrated that absorptive capacity is also best seen as an interactive process, with the use of power also influencing this process.

? DISCUSSION QUESTIONS

The following discussion questions are relevant to using this chapter in teaching exercises and discussions or for revision:

1. In what ways are different types of digital innovation (process/product/service) inter-related? Give examples to illustrate.

2. Why are some digital innovations disruptive? Give examples to illustrate.

3. Why might it be useful to recognise that the 'phases' of a digital innovation process (from the initial creation of an idea through development and on to diffusion and use) are interactive rather than linear? Give examples to illustrate.

4. What is absorptive capacity and why is it relevant to digital innovation processes?

Case questions

The following case questions relate to the linear and interactive approach to absorptive capacity. Try to fill in the tables below considering both of these topics.

1. How can the Tech Co. story be 'told' using the absorptive capacity construct?

	First implementation	Relaunch
Recognition		
Assimilation		
Application		

	First implementation	Relaunch
Knowledge exploration		
Knowledge exploitation		

1) How can we provide an explanation for the failure and the success at Tech Co. using absorptive capacity – namely what were the main problems that impeded the absorption of new external knowledge in 2001–2003, and what changed after 2004?

2) How was Tech Co. able to transform new external knowledge into innovation?

3) Could Tech Co. have avoided going through a failure in the first place and, if so, how?

 ## Additional suggested readings

Sun, P. Y., and Anderson, M. H. 2010. "An Examination of the Relationship between Absorptive Capacity and Organizational Learning, and a Proposed Integration," *International Journal of Management Reviews* (12:2), pp. 130–150.

References

Abrahamson, E., and Fairchild, G. 1999. "Management Fashion: Lifecycles, Triggers, and Collective Learning Processes,"*Administrative Science Quarterly* (44:4), pp. 708–740.

Arruda-Filho, E. J., Cabusas, J. A., and Dholakia, N. 2010. "Social Behavior and Brand Devotion among Iphone Innovators," *International Journal of Information Management* (30:6), pp. 475–480.

Christensen, C. M. 1997. *The Innovator's Dilemma*. Boston: Harvard Business School Press.

Cohen, W. M., and Levinthal, D. A. 1990. "Absorptive Capacity: A New Perspective on Learning and Innovation," *Administrative Science Quarterly* (35:1), pp. 128–152.

Cook, S. D., and Brown, J. S. 1999. "Bridging Epistemologies: The Generative Dance between Organizational Knowledge and Organizational Knowing," *Organization Science* (10:4), pp. 381–400.

Hammer, M. 1990. "Reengineering Work: Don't Automate, Obliterate," *Harvard Business Review* (68:4), pp. 104–112.

Hanseth, O., and Lyytinen, K. 2010. "Design Theory for Dynamic Complexity in Information Infrastructures: The Case of Building Internet," *Journal of Information Technology* (25:1), pp. 1–19.

Hasan, S. A., and Subhani, M. I. 2012. "Top Management's Snooping: Is Sneaking over Employees' Productivity and Job Commitment a Wise Approach?," *African Journal of Business Management* (6:14), p. 5034.

Huang, J., Henfridsson, O., Liu, M. J., and Newell, S. 2017. "Growing on Steroids: Rapidly Scaling the User Base of Digital Ventures through Digital Innovation," *MIS Quarterly* (41:1), pp. 301–314.

Lane, P. J., and Lubatkin, M. 1998. "Relative Absorptive Capacity and Interorganizational Learning," *Strategic Management Journal* (19:5), pp. 461–477.

Marabelli, M., and Newell, S. 2009. "Organizational Learning and Absorptive Capacity in Managing Erp Implementation Projects," *Proceedings of ICIS 2009*, Phoenix, AX.

Markus, M. L., and Tanis, C. 2000. "The Enterprise Systems Experience-from Adoption to Success," in *Framing the Domains of IT Management: Projecting the Future through the Past*, R.W. Zmud (ed.). Cincinnati,OH: Pinnaflex Educational Resources, Inc., pp. 173–207.

Pansardi, P. 2012. "Power to and power over: two distinct concepts of power?," Journal of Political Power (5:1), pp. 73–89.

Podsakoff, P. M., Todor, W. M., and Skov, R. 1982. "Effects of Leader Contingent and Noncontingent Reward and Punishment Behaviors on Subordinate Performance and Satisfaction," *Academy of Management Journal* (25:4), pp. 810–821.

Rogers, E. M. 1983. *Diffusion of Innovations*. London: Collier Macmillan.

Sun, P. Y., and Anderson, M. H. 2010. "An Examination of the Relationship between Absorptive Capacity and Organizational Learning, and a Proposed Integration," *International Journal of Management Reviews* (12:2), pp. 130–150.

Swanson, E. B. 1994. "Information Systems Innovation among Organizations," *Management Science* (40:9), pp. 1069–1092.

Tsai, W. 2001. "Knowledge Transfer in Intraorganizational Networks: Effects of Network Position and Absorptive Capacity on Business Unit Innovation and Performance," *Academy of Management Journal* (44:5), pp. 996–1004.

Yoo, Y. 2010. "Computing in Everyday Life: A Call for Research on Experiential Computing," *MIS Quarterly* (34:2), pp. 213–231.

Zahra, S. A., and George, G. 2002. "Absorptive Capacity: A Review, Reconceptualization, and Extension," **Academy of Management Review** (27:2), pp. 185–203.

3 ORGANISING FOR DIGITAL INNOVATION

Summary

This chapter considers how organisational designs (structure and process) have changed over time, often moving from bureaucratic structures (which can effectively support efficiency through economies of scale) to project-oriented modes of organising (which are more flexible and so can support innovation, especially digital innovation). It focuses in particular on sources of knowledge and how knowledge is acquired and shared in innovation processes and how this is influenced by organisational structures and processes. For example, new organisational forms are more 'open' both internally and externally, enabling more diverse access to knowledge sources. Openness is also important because, for an organisation to respond to changes in its environment – including sometimes, disruptive changes – it must not only be aware of what is happening beyond its boundaries, but also secure ideas and resources internally to facilitate responses to these changes. The chapter starts with a case illustration that helps to provide examples that are then drawn on in the rest of the chapter.

Learning Objectives

The learning objectives for this chapter are to:

1. Understand the principles of the bureaucratic form of organising and its advantages and disadvantages

2. Recognise how contingency theories evolved to take account of context as a determining influence on organisational forms

3. Appreciate the characteristics of non-bureaucratic organisational forms and how they support digital innovation and access to knowledge

4. Comprehend why networking is fundamentally important to digital innovation efforts

5. Be able to define the difference between open and closed innovation.

Case: CommCo – The use of social media to encourage simultaneously *univocality* and *multivocality*

CommCo is a large multinational telecommunications company distributed across 61 countries and with 90,000 employees. In 2005, CommCo decided to introduce a new organisational design. The high-level aim of this change was to save costs through enhancing employee productivity while better satisfying customers' needs by empowering employees to make more of their own decisions. In other words, CommCo wanted to be both efficient *and* flexible. To achieve this, it introduced a mobile workforce and cultural change initiative. For instance, before mobile working was introduced, field engineers whose role was to provide, update or repair services for customers would go to their offices to be given their daily work schedules by their managers. Under the mobile structure, they would simply be given a list of customers and problems and they themselves had to decide how to prioritise these based on the needs of the customer but also the importance of the customer to the business. In other words, they now managed their own work. We shall now describe the intranet and culture change initiatives in turn.

Managing different types of content on the intranet

A key issue, which emerged as CommCo developed the intranet for managing its new structure, was to cater for different communication needs at global (corporate) and local (lines of business) levels (Figure 3.1). As a result, CommCo needed to update the workforce on what was happening across the organisation to increase engagement and a 'sense of belonging', while reducing the amount of face-to-face contact as it moved towards a more mobile working model. Simultaneously, it had to allow employees to talk to each other – in much the same way as they would if they had been physically present in the workplace. It did this by providing different virtual communication channels.

More specifically, CommCo created a differentiation between two types of content on their intranet: one for that formally published by dedicated, trained communication specialists (i.e., Organisation Published Content, or OPC); the other published by users (i.e., User Generated Content, or UGC) that was explicitly geared to a more informal sharing of information and ideas. In this way, OPC was explicitly designed to provide for the formal, centralised and top-down mode of communication, which can be described as *univocal* (a single voice on the organisation strategy and operations), while UGC channels were designed to allow distributed, informal and participative communication, which can be described as *multivocal*. These two types of virtual communication mirror the traditional distinction in organisations between formal, top-down communications coming via one's line manager and informal, peer-to-peer communications that arise from daily interactions which happen when people are working in a shared physical space.

Initially, there were some tensions in using the intranet to manage the business. We next describe two tensions that were created but also how

Figure 3.1

Servers handling an organisation's digital communication
©Getty Images

CommCo overcame these tensions to allow the two modes of communication to exist simultaneously.

First, during the early stages of the rollout of the intranet, there was considerable fear on the part of some middle managers. They were afraid of losing their ability to control what was being published. This was because, in the past, many managers had worked on the assumption that power came from the control of information. This initial reluctance to engage with the intranet by some middle managers was not, however, replicated by the top management team. In particular, the CEO saw great potential in using the intranet to enhance communication among the growing mobile workforce. This drive from the top of the organisation was accompanied by business initiatives to encourage adoption of the intranet within local teams. Senior leaders did not impose cross-organisational systems but instead created a 'light' usability and navigation framework. This allowed local teams to operate independently and adjust the platform to their specific communication needs.

Second, the focus on using the intranet as a driver for the business caused tension between a central message that the organisational leaders wanted to communicate and competing demands from employees' interests and needs at the local level. This tension was managed by explicitly distinguishing between the two types of content (OPC and UGC). For example, CommCo created different colour schemes, one for OPC and one for UGC. By using different colours for each type of content, users were able to differentiate between the two and adjust their expectations about the quality and legitimacy of the sources of information.

Nurturing a communication culture for OPC and UGC

The intranet, with its social media features, was perceived to be a key part of a more ambitious programme of cultural change within CommCo. The company saw the mobile working initiative as a key enabler to modernise CommCo in terms of attracting and retaining talent and creating a more flexible workforce, so reducing the dependency on office-based work and tight managerial control. To support this transformation, CommCo created a programme to promote four core new values for the organisation: 'open', 'inspiring', 'straightforward' and 'helpful'. *Openness* was an important value closely related to the adoption of the intranet. CommCo was keen to encourage employees to have a voice and share information and ideas in a more open and transparent environment. However, it was also keen to highlight that with this power came responsibility and that all comments and views should be attributable to ensure that people did not misuse the opportunity.

The second organisational value was to be *inspiring*. This value was also closely related to the use of the new intranet platform. It was an important driver for adoption and for producing useful content and contributions. To be inspiring was key to ensure that the ideas shared would be relevant and important to the users. The third value encouraged within this new culture was to be *straightforward*. This value was promoted widely and incorporated in training manuals for all publishers responsible for OPC communication. Content and information had to be to the point and without unnecessary terminology. The fourth value was to be *helpful*. It referred to two aspects. On the one hand, it referred to the service orientation of CommCo's intranet: 'enhance user experience', 'users find what they need', 'information built around users' needs' and 'users control information consumption'. On the other hand, it was aimed at fostering users' sense of ownership and shared responsibilities when communicating on the intranet.

These values contributed to creating the cultural context necessary for the effective use of the two modes of communication in CommCo, so that employees engaged in online discussions for different purposes and were able to distinguish between personal viewpoints published by users (UGC) and facts checked by experts (OPC). Together, these two key initiatives helped CommCo to move from a traditional bureaucratic organisation to a more flexible and distributed organisation while maintaining a focus on efficiency. We draw on the case as we next describe different approaches to organisational design.

Introduction: Classic organisation design

Classic organisation and management theory assumed that is was possible to articulate a singular 'best practice' organisational design and approach to management. We introduced Taylor's Scientific Management in Chapter 1 and noted that he was focused on tight supervisory control, where workers were simply required to follow instructions and were given small tasks that had been predefined in terms of how they should be most effectively done. In terms of a more organisational level of analysis, the most well-defined classic theory of organisation comes from Max Weber in his articulation of the features and functions of the *bureaucratic* form of organising. The 2018 article by Sammi Caramela entitled 'The Management Theory of Max Weber' offers a succinct overview of this theory. Weber started his analysis by looking at different bases of authority, identifying three different bases for leaders to acquire the legitimacy needed to lead – traditional, charismatic and legal-rational. *Traditional authority* is treated as legitimate because it 'has always existed like this', often based on inheritance, as when monarchs take their position because they are the first-born (or sometimes first-born male) to the previous monarch. *Charismatic authority* is based on the personal attributes of the leader, who assumes the position because she or he is seen to have some unique (sometimes magical) qualities or personal charisma. *Legal-rational authority* is based on a system of rules that operate according to predefined principles that are assumed

Figure 3.2

Authority is a multifaceted construct
©Getty Images/iStockphoto and ©Getty Images

to create an efficient way of organising, with a person's authority based on the requirements of the particular 'office' occupied (Figure 3.2).

Weber predicted that legal-rational authority would come to dominate in organisations because it is based on a system of rules that can be designed, with the organisation then essentially working 'like clockwork'. People occupy offices with prescribed tasks and responsibilities which should ensure that things are done according to supposedly rational principles rather than the whims of either charismatic or hereditary leaders. This legal-rational basis of organisational authority underpins the bureaucratic form of organisation that Weber believed would come to dominate society. This domination would happen, he argued, from an efficiency and rationality perspective, although this form of organisation was a threat to individual freedoms because people would become mere rule-followers.

Weber listed several features that are characteristic of the 'ideal' bureaucratic form of organising, including:

1. *Specialisation:*
   Employees would specialise and so become expert in a narrow range of tasks and make decisions related to this sphere of competence based on a rational deduction of the situation at hand. Specialisation occurs at both the individual level and at the collective level, with departmental specialisation being based on a functional division of work – Research & Development (R&D), Manufacturing, Purchasing, Marketing, Sales, Logistics, Accounting, etc. The individual-level specialisation mirrors Taylor's idea of dividing work up (within these functions) into smaller tasks so that people can learn to do the tasks efficiently.

2. *Hierarchical Control:*
   Given the high level of differentiation created by specialisation, integration becomes important to ensure coordination across employees and departments – so if sales are selling 100 'widgets' per week, a company

would not want manufacturing to be producing 200 or only 50 (since there would be either a costly surplus or angry customers not getting the product they had paid for). The bureaucratic solution to this is having a chain of command, with instructions coming from the top – so that there is a single point of control – and being passed down the line – so that at each level, activities are undertaken having been predefined according to the plan. Information would then be passed back up the chain of command so that performance can be monitored to ensure that everything is working well.

3. *Formalisation:*
Since employees occupy 'offices' that are temporary, it is important that they work according to prescribed job descriptions, which lay out the roles and responsibilities and detail necessary tasks and how they should be undertaken. Rules and processes more generally are written down to ensure that there is consistency of operation. In this way, when someone leaves an office, someone else can take their place and know the tasks they will be doing because they can follow the documented rules and processes.

4. *Impersonal operation:*
The idea of a bureaucracy is that it works 'like clockwork' – things happen because they have been planned to happen and people (employees and customers) are treated in a standard way to ensure fairness and consistency. Someone provides information for another, for example, because that is in their job description. It does not matter whether they like or don't like the person they are providing with information – there are no special favours for anyone.

Dysfunctions of bureaucracy

Weber's analysis was a functional analysis – he considered the function that each of the above features of bureaucracy served. Subsequent theorists pointed to the problems or *dysfunctions* with the various characteristics of bureaucratic forms of organising. For example:

1. *Specialisation:*
One of the most famous elaborations of the idea of specialisation is attributable to Adam Smith. In his 1776 book *An Inquiry into the Nature and Causes of the Wealth of Nations*, he suggests that specialisation determines a quantitative increase in productivity at the organisational, governmental and national levels. Specialisation underpins greater skills and productivity related to individuals' micro-tasks. More recent organisational theories (we will discuss them later) criticise this approach, however, arguing that specialisation might be detrimental to workers' motivation. Very specialised workers focus on their (often repetitive) micro-tasks at the expense of gaining a holistic view of the organisation (e.g., they do not understand how

their micro-tasks fit into a bigger picture concerning over-arching organisational objectives). This was a key reason why CommCo saw the need for the OPC even while the idea was to allow workers to make more day-to-day decisions themselves and become less specialised.

2. *Hierarchical control:*
   Chester Barnard (1938) demonstrated that, while there may be a formal organisation, there is also an informal organisation that is often very different and relies on personal relationships rather than formal organisational channels. In other words, communication within an organisation rarely follows the strict lines of the hierarchy; instead, communication often follows an entirely different path, as described by the notion of the 'grapevine'. This informal communication is often much quicker and more extensive than communication through the formal hierarchy – even when the information that is flowing is false. It is this informal communication, which often bypasses the hierarchy, that 'gets things done' in organisations. As an example, if unions want to highlight a grievance, one action they can take (short of calling for a strike) is to get members to 'work to rule'. Working to rule can quickly lead to problems because much of what is actually 'done' in organisations gets done in spite of, rather than because of, the rules and formal communications. How bureaucracies actually work in practice was shown by Alvin Gouldner's famous (1954) study of a gypsum plant where he demonstrated that bureaucracy can sometimes be 'mock'. That is, while rules exist, no one actually follows them strictly, and it is because of this that the organisation actually worked in practice. Again, thinking of the CommCo case, the intranet development recognised the reality of this informal aspect of an organisation and rather than try to supress it, recognised its importance by providing the UGC channels for sharing knowledge and information between peers.

3. *Formalisation:*
   Robert Merton's (1957) work pointed to the problems of formal documentation of standard procedures. This is so because, when things change, the rules might no longer apply. But rules can take on their own momentum, morphing from being a means to an end to becoming an end in themselves, as bureaucrats apply red tape even when the rules clearly do not apply in a particular case.

4. *Impersonal operation:*
   Phillip Selznick's important (1949) study of the Tennessee Valley Authority (TVA) showed that organisations were anything but impersonal, goal-oriented entities. Rather, he showed that an organisation adapts as individuals seek to satisfy their own goals and interests, which may not be the same as the stated goals of the organisation. In this sense, rather than seeing organisations as mechanical systems,

acting like clockwork, he showed that organisations are living entities which develop cultures and informal relations and where certain patterns of action become institutionalised and taken for granted, despite not being rational from the perspective of the stated organisational goals. Organisational members, then, resist actions when they do not serve their own purposes. This he calls the *recalcitrance* problem. In the CommCo case, we saw this with the initial reaction of the middle managers who resisted the intranet because they saw it as eroding their power to control information. On top of this, Selznick noted that outside entities can also deflect from an organisation's goals, with organisations co-opting outside bodies in order to defend against these external encroachments. This he calls the problem of *external accountability*.

As we will discuss next, subsequent contingency theorists suggested the type of context in which bureaucracies might be effective – and this essentially boils down to a context where there is a very stable environment, a limited range of competitive pressures and where innovation is not so much of a pressing issue. In such a context, an organisation can continue to do what it has always done and bureaucracies can support this. The implication is that bureaucracy is not appropriate for many contemporary organisations that are working in a context characterised by high levels of global competition and rapid change. Organisations must therefore continuously respond to changing circumstances. Certainly, it was the case with CommCo that the reason it was changing its organisational design was to allow it to be more flexible and responsive to customer demands.

Given the limitations of bureaucratic forms of organising, there have been some who have predicted that this form of organising would disappear (Heckscher and Donnellon, 1994) and new organisational forms, which we discuss more comprehensively below, would come to dominate (Peters, 1992). In reality, bureaucratic structuring is still very prevalent (Alvesson and Deetz, 2006) and can serve an enabling role in terms of getting work done efficiently, even while it may be experienced as coercive by employees (Adler, 2012) and, as we will see, limiting for organisations in driving innovation. This is perhaps not surprising since formal control is still needed in organisations and indeed may become even more important as organisations become more globally distributed. Interestingly, we see in the CommCo case that it had maintained some degree of central control with its OPC. However, here, this was balanced with the UGC providing both central and more decentralised communication (and so control) simultaneously.

Contingency theories

According to contingency theories, there is no optimum way to organise a company but instead, because every company is different, best fit is viewed as a superior approach. One such early example (in practice) of contingency theories is represented by the work of Joan Woodward, who, in the 1950s, demonstrated that the use of specific technologies determined how people organise companies. Having studied 100 manufacturing companies in the UK, she found that three types of productions (i.e., technologies) – small batch and unit production; large batch and mass production; and continuous process production – were associated with different management structures, distribution of work, responsibilities, levels of accountability and skills of workers. We can also point towards the seminal work of Henry Mintzberg on organisational structure, which outlined five structural types through which organisations (attempt to) enact their strategy (Mintzberg, 1980).

While studies have demonstrated that organisations do not necessarily operate in the 'clockwork' fashion assumed by the bureaucratic ideal, perhaps a more important problem with bureaucracy that studies began to highlight was its inability to adapt to changes in the environment. One of the early studies to demonstrate this was conducted by Burns and Stalker (1961). They looked at a small sample of firms that were operating in an environment where technology was changing rapidly – what would today be classified as *disruptive innovation* (Christensen, 1997). For example, new technology has made mechanical weighing scales obsolete in many contexts because electronic scales have superior accuracy. As another example, digital cameras made film-based cameras and more importantly camera film itself virtually extinct. This created major problems for the US company Eastman Kodak (normally referred to as Kodak) that had relied on a 'razor blade' strategy for growth (i.e., selling cameras very cheaply because it was the sale of rolls of film where its profits were made, just as the disposable blades create the value for sales of razors). Kodak found it difficult to respond to this new era of digital photography, even though the company had invested in and developed some of the early breakthroughs in this space (e.g., Lucas and Goh, 2009).

In contexts of disruptive innovation, companies must develop new knowledge and expertise, even while often they also have to continue to develop and sell 'old' technology. This is so because an existing customer base does not change overnight and so still needs to be catered for and indeed will remain the major source of revenue during the early period of change. The 'old' part of a company is often very powerful in such situations, with vested interests and intransigent assumptions about how value is created, making it difficult for those working on the new technologies to be heard. For this reason, many established firms struggle in periods of disruptive innovation, as did Kodak, which filed for Chapter 11 bankruptcy protection in January 2012. Kodak emerged from this as a company focused on the corporate digital-imaging market rather than providing cameras and more importantly rolls of film for home and professional photographers.

In the study by Burns and Stalker, they observed that some firms were able to adapt relatively easily to this type of changing environment, while other firms were much less successful. They also observed that the firms in their sample were organised differently – organisational forms that they described as mechanistic or organic. The *mechanistic* organisations, which were essentially bureaucratically organised, having tall hierarchies, considerable specialisation and formalisation, for example, were the firms that were unable to adapt to the changing environment. The *organic* organisations, which essentially had the opposite characteristics – flat hierarchies, low specialisation, little formalisation etc. – were much more able to adapt. This, then, is a contingency theory (rather than the one 'best practice' idea characteristic of traditional theories about how to organise). In this light, an optimal, or at least 'good', organisational form depends on (i.e., is contingent upon) something else. In Burns and Stalker's study, the contingent variable is the environment and whether this is stable or dynamic. Their conclusion was that mechanistic, bureaucratic forms of organising are good for when the environment is stable, but not for when the environment is dynamic, because such structures make it difficult to introduce change and innovate. The CommCo case is an example where the organisation recognised that its old bureaucratic structure was no longer effective; this was why it had introduced the mobile initiative and started the process of culture change to become a flatter, leaner and more flexible organisation (Figure 3.3).

Figure 3.3

Different ways to view how organisational components can be combined
©GETTY

🌑. **Key Concepts: Classic Organisational Design**

Bureaucracies:

- Weber's traditional, charismatic and legal-rational forms of authority, featuring:
- Specialisation – functions and individual jobs characterised by limited scope of tasks
- Hierarchical control – centralised control of functions and individuals
- Formalisation – rules, procedures, job descriptions and so forth all written down to be followed precisely
- Impersonal operations – individuals undertake tasks because it is in the job description, not because of any social networks and relationships

Dysfunctions of bureaucracies:

- Informal organisations – social networks and relationships exist and influence what happens despite attempts to override this with rule-following
- Red tape – rules become ends in themselves rather than a means to an end

Contingency theory:

- No one 'best practice' organisational design
- Bureaucratic or organic organisations 'best' depending on stability of the environment

Bureaucracies: Digital innovation and change

It is helpful to think about some of the reasons why bureaucratic forms of organising can be poor for stimulating digital innovation and change. Some of these reasons include the following:

1. *Communication:*
   Communication can be slow because of the many layers in the hierarchy that have to be gone through. Moreover, since a communication has to pass through multiple people before it reaches its destination, it can often be distorted (either accidentally because of misunderstanding or purposefully to serve political ends). Even aside from these problems, with centralised communication structures, decision-making tends to end up at the top, leading to executive information overload and their resulting maladaptive responses such as omissions, errors and approximations (Katz and Kahn, 1978). Clearly, CommCo had recognised the limits of relying on centralised communication structures, even though initially middle managers in particular had resisted the move to allow employees to communicate across the organisation more freely using UGC.

2. *Control:*

High levels of specialisation characteristic of bureaucratic organisations mean that there must be high coordination and formal control to ensure that the 'pieces' fit together. Bureaucracies achieve this control in a number of ways, including standardisation of activities, closeness of supervision and hierarchical authority. However, these are rigid systems of control that make it difficult to respond to changing circumstances. What is required is an organisational form that is able to be more responsive because it allows employees to modify rules and procedures as they come to recognise changed circumstances. This is sometimes described as informal control (rather than the formal control of a bureaucracy) and includes an emphasis on socialisation so that employees internalise the norms and values of their organisation and so will adapt to its interests rather than blindly following rules. The culture initiative in CommCo was aimed at creating such shared norms and values, even while the UGC allowed users to freely communicate with each other.

3. *Centralised decision-making:*

In bureaucracies, decisions are made at the top of the hierarchy and then passed down the chain of command and implemented at different levels. In this sense, decision-making is centralised. However, often it is those on the ground who know how the environment is changing and how the organisation might be better able to respond to such changing circumstances; those at the top may be less aware of the need to change, especially since communication upwards is often filtered. In CommCo, allowing employees to manage their own work schedules (rather than their managers) was clearly a move to decentralise decision-making. This illustrates the need for flatter and more decentralised organisational structures, a point to which we now turn (Figure 3.4).

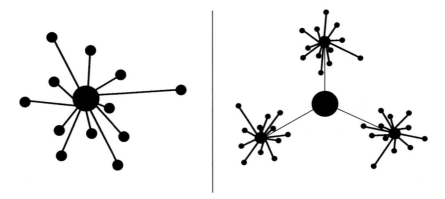

Figure 3.4

Centralisation versus decentralisation

New forms of organising

While there are certainly still some (or parts of some) organisations that remain heavily bureaucratic, it's also the case that many organisations have changed since the latter part of the twentieth century, and many of these changes have increased their ability to adapt and innovate and operate more globally and with less formal hierarchy and control (Friedman, 2006). Some key aspects of these changes include the following:

1. *Flatter structures and more decentralised decision-making*

While hierarchy can help with coordination and resolution of conflicts (Galbraith, 1977), it can also make it difficult to change (Burns and Stalker, 1961). Indeed, this consequent structural inertia was the basis for Hannan and Freeman's (1977) argument that, rather than organisations adapting to their environment, a stronger dynamic explaining the success of organisations is that this follows a process of 'natural selection' (similar to Darwin's theory of evolution). To briefly summarise their idea, which they call *population ecology*, organisations compete for resources in a particular niche. When a new niche emerges, organisations will begin to compete in this niche and some organisational models will appear to be more successful so that others will copy this. Over time, as the density of organisations in this niche increases, resources will become scarce (because all the organisations are competing for the same resources), so that eventually the number of competing organisations will decline because the niche cannot sustain them all. This means that some organisations will fail and go out of business. Organisations that are reliable and accountable are most likely to be successful in the early period of niche emergence – because, as the proponents of bureaucratic and hierarchical organisation remind us, these structural characteristics can be very efficient. However, these organisations are also likely to be difficult to change and so ultimately it is these inertial forces that will lead to them being selected out.

In place of tall hierarchies (as in Organisation A in Figure 3.5), flatter structures (Organisation B) allow more decentralised decision-making,

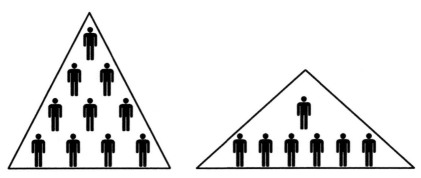

Figure 3.5

Hierarchical versus 'flat' organisations

thereby reducing the burden on top executives and allowing quicker and more responsive decision-making, nearer to the place and time where the decision is needed. Decentralised decision-making can be more effective, especially in situations where the solution is not simply dependent on pooling independent pieces of information. In addition, it leads to greater satisfaction among those involved. This was illustrated in a series of classic studies by Shaw (1982), who used laboratory experiments to compare decision-making success and staff member satisfaction within various structural arrangements. For example, the individuals in a five-person team might each be given a piece of information that is needed to solve a problem. Some teams can communicate with whomever they want (i.e., all channels are open so that the structure is decentralised), while other teams can only communicate 'up' the chain of command. In the latter centralised, hierarchical case, a person would have to communicate their information to the next person, who must then communicate their information plus the information acquired from the person below to the next person and so on until all the information eventually reaches the person at the top of the chain. The person at the top would then have to put all the information together and find a solution or make a decision. Findings from this type of study show that, while the centralised structure is quicker at solving problems that are simple, the decentralised structure is much better at solving complex problems – because more people can offer suggestions and debate can stimulate ideas for solutions. Furthermore, in centralised structures, while the central person or people may be satisfied with the process, the rest of the team may well be dissatisfied; in decentralised structures, all members tend to be more satisfied. Cummings and Cross (2003) confirmed these results in actual work teams, finding that hierarchical structures were negatively associated with group communication and performance.

CommCo's mobile initiative was aimed at decentralising the structure, so that employees would become less reliant on their manager for making decisions about their work schedules. In this sense, CommCo was moving away from traditional forms of bureaucratic control: output control and process control. Output control sets targets and monitors the outputs of work against this target; process control defines how the work is done so that people have to follow this process (the assembly line is a useful way of thinking about this type of control). In flatter organisations, emphasis is often on normative control, where a set of shared norms and a shared sense of identity mean that people take actions consistent with the organisational goals without the need for centralised hierarchical control. Ouchi (1981) refers to this as *clan control*. The four elements of the new culture at CommCo are an example of an attempt to implement such clan control rather than rely on bureaucratic control.

While clan (or normative) control may be increasingly important in many settings, this does not rule out the need for more formal hierarchical control. Indeed, Brenner and Ambos (2013) showed that efforts to instil shared norms across multinational organisations, for example through train-ing and culture building, which they described as social forms of control, were an important precursor to being able to use more formal hierarchical controls. Without this initial creation of shared norms, they argue, the for-mal controls would not be accepted as legitimate and so would not be effec-tive. The fact that OPC remained an important source of centralised information in CommCo is illustrative of this continued importance of hier-archical control. This indicates that some form of hierarchical control remains important in many organisations, even if it is often now used along-side more social and normative forms of control.

From the above, it is clear that the various types of control have strengths and weaknesses. As we pointed out earlier in the book, tight control (Taylor, Chapter 1) and intense specialisation (Smith, this chapter) can lead to increased productivity. This type of control is currently adopted in large manufacturing companies where efficiencies created through economies of scale and intense specialisation combine to enable mass production at mini-mum costs. In these organisations, routine-based jobs such as those related to assembly line work may create alienation because workers often do not realise the extent to which their micro-tasks contribute to a complex finished product. Therefore, bureaucratic forms of control emerge to ensure that poorly motivated workers show up at work and execute their tasks in line with their job description.

On the other hand, organisations where culture/clan control is prevalent are those (flat-structured) innovative firms where most organisational actors are required to have task-switching abilities (doing different types of jobs requiring a mix of skills) and where individual initiative is rewarded (rather than punished, as could have happened in the typical 'Tayloristic' organisa-tion at the beginning of the twentieth century). These 'flatter' organisations focus on innovation rather than efficiencies and economies of scale and good examples are high-tech companies, start-ups and the like. For instance, in a 2010 interview, Steve Jobs, at that time CEO of Apple, explained how his company's culture allows everyone to contribute to the company's success with creative ideas and personal initiatives. He defined Apple as 'the largest start-up company in the world'. Clearly, organisations such as Apple need constant innovation and therefore there is little room for 'siloed' jobs (where workers focus on a single task) that are tightly controlled.

One last thing worth noting: because of the increased global competitive environment, nowadays organisations are required to constantly innovate while trying to be efficient (e.g., delivering novel products and services at a minimum cost). Therefore, some organisations focus on bureaucratic control at the bottom of the pyramid (e.g., production line) and promote a clan/cultural type of control at other (upper) levels and/or in specific business units such as R&D and marketing, where novel ideas generally emerge. Next we discuss, specifically, the relevance of business units within many contem-porary organisations.

2. *Business units*

Rather than being a single large entity, organisations (at least larger organisations) are often divided into smaller units or divisions. The rationale is that having multiple units allows each to specialise and adapt to the particular contingencies of its task and market environment (Chandler, 1962; Lawrence and Lorsch, 1967). This is important because, as organisations grow more complex (e.g., developing more complex technologies, operating across multiple markets, sometimes globally distributed, including considerable product and service variety), it becomes very difficult for managers to understand and be in control of everything that is going on. As we probably all appreciate, human beings – and that includes managers – have cognitive limits (Simon, 1991). Moreover, as organisations diversify, their different products and services require different approaches to technology development, marketing, after-sales service and so on. As an illustration, if you buy an expensive luxury car, you expect a different level of service to when you buy a very basic, much cheaper model. You also expect the technology to be more leading-edge when you buy the luxury brand. Similarly, even businesses such as McDonald's have a slightly different menu in different locations in recognition of different tastes and consumption habits. Compare, for example, a McDonald's in New York City to one in Hong Kong. Dividing into semi-autonomous business units allows an organisation to differentiate in order to provide these different types of product and service to different market niches.

Creating different divisions for different products and markets is one approach to dividing up an organisation. Another approach is *modularisation* (Baldwin and Clark, 2000), which makes divisions by task rather than product or region. The idea of modularisation is to group together into separate units those tasks that are highly interdependent while minimising the dependencies between the units, so that each unit can make decisions in a self-contained way (Siggelkow and Rivkin, 2005). However, there is obviously a trade-off since creating too many independent units risks losing out on potential synergies across the units. Furthemore, the degree of modularisation will also depend on the extent of interdependence between the various tasks in which an organisation is involved – it is more difficult to modularise in some organisations because the tasks are highly inter-related.

Consider Table 3.1. Imagine, for instance, there are nine tasks, each being more or less interdependent with each other task – with the x denoting that there is a high interdependency between two tasks.

We have three clear task clusters – within a particular cluster, tasks are interdependent, but between clusters they are fairly independent. Modularisation in this example would be simple – we would divide the organisation into three: the yellow, blue and green modules or business units. This modular approach to designing organisational structures is tightly tied

Table 3.1 An example of simple modularisation

Tasks	1	2	3	4	5	6	7	8	9
1		X	X						
2	X		X						
3	X	X							
4					X	X			
5				X		X			
6				X	X				
7								X	X
8							X		X
9							X	X	

Table 3.2 A more complex example of modularisation

Tasks	1	2	3	4	5	6	7	8	9
1							X	X	
2				X					X
3		X			X				
4	X					X			
5			X					X	
6			X	X					
7						X			X
8	X						X		
9		X				X			

up with modular approaches to product design. In modular product design, a final product is divided into distinct parts and standards are agreed about how the parts will fit together. Once these standards are agreed, each part can then be developed and modified independently as long as the standards for working with other parts are not changed (e.g., Dell computers adopted this strategy, allowing customers to build their own computer because they could select the particular specification of the component parts that suited their needs, with all parts being interchangeable).

Now, imagine an organisation where interdependencies between tasks are less uniform. In Table 3.2, tasks are not decomposable into neat modules. In such an organisation, business units might be created, but there would be a need to maintain a much higher level of coordination across the units to ensure that they were communicating, sharing information and liaising on decisions so that the interdependencies on tasks outside the division were taken into account. However, this coordination does not necessarily have to involve additional structural layers, as in traditional hierarchies; instead, in today's flatter organisations, it is more likely to involve the inclusion of more supervisory units that can provide coordination across units (Zhou, 2013), as discussed in the previous section.

3. *Project forms*

Zhou's reminder that hierarchical control does not have to involve adding layers to the communication chain is also relevant to another change relating to the way organisations are often structured today; that is, rather than focus on coordination through vertical control (up and down the hierarchy), the emphasis is on horizontal coordination – across different functional departments. One solution to this horizontal coordination is to create a *matrix* structure, where the functions remain but a horizontal project or product structure is superimposed. For example, with a horizontal product structure, a product manager would be responsible for the whole life cycle of the product, including all the different functional tasks that are involved. In such circumstances, the product manager brings in resources from the functions and these functional representatives are responsible for this particular product line and coordinate their particular functional requirements with other functional experts also involved. So, in a matrix structure, functions do not simply pass on their work to the next function but rather work together to identify solutions for a particular product that are best from an overall process perspective, not simply from their own, narrow, functional perspective.

For example, R&D may come up with a new design for a product that exploits digital technological developments and potentially creates a better product. When this new design is passed to Manufacturing, they may baulk at the design because it might require a totally different manufacturing set-up, which might be infeasible within the time scale and resources provided. Manufacturing would then throw the design back to R&D and ask them to amend it to take into account manufacturing resources. This could happen many times before a compromise is finally reached. In a matrix structure, theoretically, R&D and Manufacturing (plus all the other functions) work together from the start to ensure that their different interests and capabilities are taken into consideration. In reality, having both functional and product demands is likely to lead to inconsistencies, such that individuals face conflicting demands on their time (Boettger and Greer, 1994).

Matrix structures are found in many different organisations and can overcome the problem of functional silos. Kuprenas (2003), for example, describes the introduction of a matrix structure in the City of Los Angeles, Bureau of Engineering. The study demonstrated that there can be problems associated with this kind of structure. Power dynamics, for example, can become more intense in matrix structures because a horizontal control structure is superimposed upon the hierarchical functional structure, meaning two sets of managers are vying for the same resources. Despite these problems, in this case at least, the performance of the organisation with respect to project delivery was improved. Indeed, matrix structures are very common in organisations where projects – whether projects to develop innovations in a product or service, or consultancy projects – are important.

A step beyond a matrix structure is to create a full process structure – where the functional/hierarchical lines are dissolved and multifunctional teams work together throughout the life cycle of a product or service (albeit a particular type of expertise may be more or less important during different stages of the life cycle). We discuss this more fully in Chapter 5 when we consider teams and projects.

An important point to remember as organisations move from vertically structured, hierarchical forms to more horizontally, matrix and process structured forms is that such changes place new demands on the communication channels within the organisation. As we have noted, in vertically structured organisations, there is a single line of command and communication flows up and down the hierarchy (at least formally). We have discussed how sometimes this flow of communication is not very effective. For example, individuals might restrict the flow of negative information upwards given concerns that this would make them and/or their department look as if they are not managing well. Therefore, it is often the case that alongside such formal flows of communication, there are informal flows, such as 'the grapevine', where information can actually flow very quickly, regardless of whether it is true. However, in vertically structured organisations, there is at least a clear and single flow of formal communication, which means that it is easier to ensure consistency in formal messages. There is one voice – or *univocal* communication – as with the OPC in the CommCo case.

In horizontally structured organisations, where communication flows horizontally as well as vertically, maintaining a single voice is more difficult and there will be multiple voices – referred to as a *multivocal* environment – that might not always be consistent, potentially creating tensions between what is communicated by different individuals or teams. This is especially so with the advent of email and social media inside an organisation that can allow greater reach of communication across an organisation given that the traditional grapevine relied on word of mouth (Huang et al., 2013; Baptista et al., 2017). CommCo provides a good example, where the use of digital technology included both univocality and multivocality in order to get the benefits of both types of communication. Moreover, this ability to communicate through digital technology has enabled employees to be physically distributed as well – people can be working on a joint task or project but not co-located either because they live in different places or simply because – as increasing numbers now do – they work from home. This, however, can create issues in achieving common understanding. These communication tensions, and how they can be managed, in a flatter and more process-oriented organisation are illustrated in the CommCo case.

4. *Networking*

The emphasis on new forms of organising discussed so far has still kept everything 'in-house', so to speak – the tasks are just organised differently to allow more flexibility and increased innovation. However, another type of change is that organisations, rather than trying to innovate in-house, now often work in collaborative partnerships in what has been described as an open innovation approach (Chesbrough, 2003). Digital technologies can facilitate this increased organisational openness. We discuss open innovation more fully in Chapter 6. Suffice to say here, new organisational forms increasingly rely on working collaboratively with other organisations so that alliance building and nurturing become key organisational competencies.

Owen-Smith et al. (2002) have suggested that there are crucial national differences in innovation trajectories that can be explained by differences in organisational capabilities to develop organisational alliances – capabilities they defined as relational and integrative. They defined *integrative capabilities* as the ability to collaborate with diverse organisations and *relational capabilities* as the ability to integrate knowledge across diverse communities. These capabilities, they argue, are variable across national contexts; in other words, in some countries, relational and integrative capabilities are better developed because the institutional context is more supportive of such capabilities. For example, relational capabilities are better in countries where academic and industry connections are stronger, such as in the USA, and worse in countries like the UK where universities have tended to have a greater academic ('ivory-tower') focus.

Another factor that affects national differences in the connections that are important for innovation is a nation's level of *corporatism* – a term introduced by Vasudeva et al. (2013). They found that a nation's level of corporatism influences partner selection in alliance formation as well as the manner in which knowledge is acquired from a partner. Corporatism relates to the concentration and ongoing influence of a country's interest groups (e.g., trade associations, business confederations, industry associations) and is a product of the institutional environment (i.e., the taken-for-granted assumptions about how, in this case, the business environment operates). In highly corporatist settings (such as Germany and Japan), interest groups are stable and highly organised and participate in the setting and implementation of economic policy, based on principles of consensus and cohesion that create highly coordinated and collaborative business networks (Whitley, 2000). In less corporatist settings (such as the UK and USA), interest groups tend to be more narrowly focused and insular, arising to address particular issues but then dissolving and not resulting in strong and dense long-term collaborative networks. Based on such observed national institutional differences, Vasudeva et al. (2013) hypothesised (and confirmed) that in high corporatist contexts, alliance partners would be sought that could help a firm build its connections within the existing networks of collaboration, with knowledge considerations playing a secondary role and, indeed, with knowledge acquisition being seen as a two-way mutual process. On the other hand, in less corporatist contexts, partners would be sought in a much more exploitative fashion, focusing on the knowledge that could be acquired from the partner in as short a period as possible, and with attempts to actively protect what the partner could learn from the acquirer. Such differences indicate why the same government policy for stimulating innovation may not work in all contexts (Swan et al., 2010); policies that encourage collaboration between partners not used to collaborating may not be very successful in a less corporatist institutional environment.

So, working in networks with alliance partners can support innovation in various ways and this may vary across different institutional contexts. However, at the same time, research has shown how these networks themselves can become stagnant. Therefore, in relation to network evolution, while partnering with other organisations can facilitate broader access to knowledge (and other resources) for innovation, it is also the case that firms tend to develop alliances with familiar partners, so that over time the networks can become a stabilising (inertial) force rather than a source of variety. However, Rosenkopf and Padula (2008) identified how the involvement of new entrants into established networks helped to create new clusters that could lead to fundamental network change and so be a source of new ideas. This indicates that it is not simply that firms can benefit in being more open in their approach to innovation but that they can also benefit from exposing themselves to new network partners in order to ensure that their network remains adaptable.

Aligning old and new organisational forms

Much of the discussion in this chapter suggests that an organisation must choose between a more bureaucratic structure and a more organic structure. However, in reality, an organisation can be ambidextrous (Duncan, 1976) (see Chapter 1). This means it can be both things – both efficient and innovative/flexible – at the same time. We will not discuss this further in this chapter, but very obviously CommCo can be described as an example of an ambidextrous organisation at least in relation to its communication approach – it maintained some central control of its communication while also providing channels where there was no managerial oversight of what was communicated. Huang et al. (2015) explicitly refer to this as communicational ambidexterity. The two key drivers to this were the use of digital technology to enable different types of communication combined with a programme to instil the type of culture that would allow it to operate flexibly while still maintaining some central control. This reminds us that, while organisations may want to unleash the innovative capabilities of employees, they will do this within the confines of strategic intent.

After all, organisations exist for a specific purpose – they are more than just a collection of people. So, even though today innovation may be important for many organisations, the innovation is focused on what each organisation has been set up to achieve.

 Key Concepts: New Organisational Design

Different ways to ensure organisational control:
Output control – control by defining output targets
Process control – control by defining the process by which work must be completed
Clan or normative control – control by developing common norms and social bonds

New organisational forms:
Business units – dividing up the organisation into semi-autonomous business units that can focus on a particular product range or region
Modularisation – dividing up by tasks that are interdependent and making product components interchangeable
Project and matrix structure – having a horizontal system of communication and control as well as a vertical system
Networks and open innovation – recognising that working with external partners can help an organisation to be more flexible and innovative

Institutional influences on open innovation:
Integrative and relational capabilities – organisational capabilities to network with other organisations based on their ability to develop partnerships and share knowledge
Corporatism – the concentration of a country's interest groups that influences how far organisations are likely to work cooperatively together versus competitively

Communicational ambidexterity:
Univocality and *multivocality* – which allow a single voice and multiple voices to be heard simultaneously that potentially facilitate the joint achievement of hierarchical and horizontal communication

Conclusions

In this chapter, we have examined how organisations have changed in recent times and we have also considered explanations for these changes. This has allowed us to explore the advantages of non-bureaucratic organisational forms of the late twentieth and early twenty-first centuries, especially in contexts where innovation is important. We have also seen how digital technology has influenced these organisational changes, as illustrated in the

CommCo case. However, while we have demonstrated that organisations have undergone some significant changes in the ways they are structured and organised, we should not take this too far – many organisations still attempt to impose control on their now very distributed activities and the old power dynamics that allowed some groups to continue to prosper at the expense of others (not the ideal meritocracy that is sometimes declared) are still very much with us. In the next chapter, we continue with this theme by looking at how strategy is an inherently political process that is emergent and on-going rather than simply a planned and rational response to the external contingencies that are objectively perceived.

? DISCUSSION QUESTIONS

The following discussion questions are relevant to using this chapter in teaching exercises and discussions or for revision:

1. What are the advantages and disadvantages of bureaucratic forms of organising?

2. What is meant by a contingency theory and what is an example of a contingency that will affect organisational structure?

3. What are some features of organisational structure that support innovation and how can digital technology support such structures?

4. How can these same features also reduce efficiency and how can digital technology help overcome these?

💬 Case questions

The following case questions might also be relevant to using this chapter in teaching exercises and discussions:

1. What evidence is there that CommCo can be described as an ambidextrous organisation?

2. What characteristics of its use of the intranet allows CommCo to provide a single organisational message while allowing people to have their own say within the organisation?

3. Why did CommCo need to change culture as well as the structure of its communication channels as it tried to be both flexible and efficient?

4. How do you think the specific values of 'open', 'inspiring', 'straightforward' and 'helpful' were important in CommCo's move to become a more flexible organisation?

Additional suggested readings

Chesbrough, H. and Garman, A. (2009). How open innovation can help you cope in lean times, HBR, December.

Whelan, E., Parise, S., deValk, J. and Aalbers, R. (2011). Creating employee networks that deliver open innovation. *MIT Sloan Management Review* 53:1, 37–40.

References

Adler, P. S. 2012. "Perspective – the Sociological Ambivalence of Bureaucracy: From Weber Via Gouldner to Marx," *Organization Science* (23:1), pp. 244–266.

Alvesson, M., and Deetz, S. 2006. "Critical Theory and Postmodernism Approaches to Organizational Studies," in *The Sage Handbook of Organization Studies*, S.R. Clegg, C. Hardy, B.S. Lawrence and W.R. Nord (eds.). London, Thousand Oaks, New Delhi: Sage, p. 255.

Baldwin, C. Y., and Clark, K. B. 2000. *Design Rules: The Power of Modularity*. Cambridge, MA: MIT Press.

Baptista, J., Wilson, A. D., Galliers, R. D., and Bynghall, S. 2017. "Social Media and the Emergence of Reflexiveness as a New Capability for Open Strategy," *Long Range Planning* (50:3), pp. 322–336.

Barnard, C. I. 1938. *The Functions of the Executive*. Cambridge, Massachusetts: Harvard University Press.

Boettger, R. D., and Greer, C. R. 1994. "On the Wisdom of Rewarding a While Hoping for B," *Organization Science* (5:4), pp. 569–582.

Brenner, B., and Ambos, B. 2013. "A Question of Legitimacy? A Dynamic Perspective on Multinational Firm Control," *Organization Science* (24:3), pp. 773–795.

Burns, T. E., and Stalker, G. M. 1961. *The Management of Innovation*. London: Tavistock.

Caramela, S. 2018. *The Management Theory of Max Weber*. Business.com, Available at: https://www.business.com/articles/management-theory-of-max-weber/

Chandler, A. D. 1962. *Strategy and Structure: Chapters in the History of the American Enterprise*. Cambridge, MA: MIT Press.

Chesbrough, H. W. 2003. "The Era of Open Innovation," *MIT Sloan Management Review* (44:3), pp. 35–41.

Christensen, C. M. 1997. *The Innovator's Dilemma*. Boston: Harvard Business School Press.

Cummings, J. N., and Cross, R. 2003. "Structural Properties of Work Groups and Their Consequences for Performance," *Social Networks* (25:3), pp. 197–210.

Duncan, R. 1976. "The Ambidextrous Organizations: Designing Dual Structures for Innovation," in *The Management of Organization*, R.H. Killman, L.R. Pondy and D. Sleven (eds.). New York: North Holland, pp. 167–188.

Friedman, T. L. 2006. *The World Is Flat: The Globalized World in the Twenty-First Century*. London: Penguin.

Galbraith, J. R. 1977. "Organization Design: An Information Processing View," *Organizational Effectiveness Center and School* (21:8), pp. 21–26.

Gouldner, A. W. 1954. *Patterns of Industrial Bureaucracy*. New York: Free Press.

Hannan, M. T., and Freeman, J. 1977. "The Population Ecology of Organizations," *American Journal of Sociology* (82:5), pp. 929–964.

Heckscher, C., and Donnellon, A. 1994. *The Post Bureaucratic Organization: New Perspectives on Organizational Change*. Thousand Oaks, CA: Sage.

Huang, J., Baptista, J., and Galliers, R. D. 2013. "Reconceptualizing Rhetorical Practices in Organizations: The Impact of Social Media on Internal Communications," *Information & Management* (50:2), pp. 112–124.

Huang, J., Baptista, J., and Newell, S. 2015. "Communicational ambidexterity as a new capability to manage social media communication within organizations," *The Journal of Strategic Information Systems* (24:2), pp. 49–64.

Katz, D., and Kahn, R. L. 1978. *The Social Psychology of Organizations*. New York: Wiley.

Kuprenas, J. A. 2003. "Implementation and Performance of a Matrix Organization Structure," *International Journal of Project Management* (21:1), pp. 51–62.

Lawrence, P. R., and Lorsch, J. W. 1967. "Differentiation and Integration in Complex Organizations," *Administrative Science Quarterly* (12:1), pp. 1–47.

Lucas, H. C., and Goh, J. M. 2009. "Disruptive Technology: How Kodak Missed the Digital Photography Revolution," *The Journal of Strategic Information Systems* (18:1), pp. 46–55.

Merton, R. K. 1957. "Priorities in Scientific Discovery: A Chapter in the Sociology of Science," *American Sociological Review* (22:6), pp. 635–659.

Mintzberg, H. 1980. "Structure in 5's: A Synthesis of the Research on Organization Design," *Management Science* (26:3), pp. 322–341.

Ouchi, W. 1981. "Theory Z: How American Business Can Meet the Japanese Challenge," *Business Horizons* (24:6), pp. 82–83.

Owen-Smith, J., Riccaboni, M., Pammolli, F., and Powell, W. W. 2002. "A Comparison of Us and European University-Industry Relations in the Life Sciences," *Management Science* (48:1), pp. 24–43.

Peters, T. 1992. *Liberation Management.* New York: Knopf.

Rosenkopf, L., and Padula, G. 2008. "Investigating the Microstructure of Network Evolution: Alliance Formation in the Mobile Communications Industry," *Organization Science* (19:5), pp. 669–687.

Selznick, P. 1949. *TVA and the Grass Roots: A Study of Politics and Organization*. Berkley, Los Angeles, London: University of California Press.

Shaw, M. L. 1982. "Attending to Multiple Sources of Information: I. The Integration of Information in Decision Making," *Cognitive Psychology* (14:3), pp. 353–409.

Siggelkow, N., and Rivkin, J. W. 2005. "Speed and Search: Designing Organizations for Turbulence and Complexity," *Organization Science* (16:2), pp. 101–122.

Simon, H. A. 1991. "Bounded Rationality and Organizational Learning," *Organization Science* (2:1), pp. 125–134.

Swan, J., Bresnen, M., Robertson, M., Newell, S., and Dopson, S. 2010. "When Policy Meets Practice: Colliding Logics and the Challenges of 'Mode 2'initiatives in the Translation of Academic Knowledge," *Organization Studies (31:9–10), pp. 1311–1340.*

Vasudeva, G., Zaheer, A., and Hernandez, E. 2013. "The Embeddedness of Networks: Institutions, Structural Holes, and Innovativeness in the Fuel Cell Industry," *Organization Science* (24:3), pp. 645–663.

Whitley, R. 2000. "The Institutional Structuring of Innovation Strategies: Business Systems, Firm Types and Patterns of Technical Change in Different Market Economies," *Organization Studies* (21:5), pp. 855–886.

Zhou, Y. M. 2013. "Designing for Complexity: Using Divisions and Hierarchy to Manage Complex Tasks," *Organization Science* (24:2), pp. 339–355.

4 STRATEGISING FOR DIGITAL INNOVATION

Summary

This chapter considers how organisations *strategise* in relation to digital innovation. The traditional view of strategy is that an organisation needs to plan ahead for what new things it wants to do, whether this is developing a new product, service or organisational process. This planned view of strategy is helpful in the sense that organisations do realistically need to develop plans about how they are going to operate and compete in the future. However, in practice, plans often do not work out and so other views have been developed that recognise the emergent and dynamic nature of strategy. This is most commonly referred to as *strategising*, which reflects the everyday 'doing' of strategy in organisations. We explore both of these views in this chapter while recognising the importance of strategy and strategising that simultaneously exploits and explores knowledge, considering aspects of digital strategising in passing.

Learning Objectives

The learning objectives for this chapter are to:

1. Understand the basic elements of a planned strategy
2. Recognise that not all strategy can be planned and that emergent features of strategy-making are important
3. Discuss the main elements of a strategy-as-practice perspective
4. Understand how knowledge and knowing link to planned versus emergent ideas of strategy and strategising
5. Be aware of the role of digital technology in strategising for innovation from a knowledge perspective.

Case: ChinaTicketCo – event ticketing in China

The live performance segment of China's entertainment industry has expanded rapidly in the last 10–15 years. This case describes the role of one ticketing firm – ChinaTicketCo – and how it has become a leader in this market sector. Starting from nothing in 2004, by 2011 the firm had a workforce of 550 with 32 branch networks and was handling over 10 million tickets annually. ChinaTicketCo is also a trendsetter in that it has fundamentally changed the practice of ticketing, as we describe next.

Phase one: Ticketing as a transactional practice

In the conventional pattern of ticketing practices in China, ticketing agencies were state-owned enterprises (SOE). SOE were, and still are in some industries, a dominant form of organisation in the country, where the state controls the business. In this initial period, an event would be initiated by an event organiser. The event organiser would then coordinate all the other involved parties (the venue, the performing artists and the media if relevant). The event organiser would also work with the ticketing company which is the main link to the audience. After the dates and venue are finalised, what is of most concern to the event organiser and the performing artist(s) is the number of ticket sales. Once a ticketing company is selected and signs the contract with the event organiser, it becomes the 'tier-one' agent, responsible for producing and distributing tickets to its distribution network of 'tier-two' agents (e.g., small box offices, travel agents, hotels' travel desks and, often, its competitors). For most people, the ticketing agency will not be particularly visible unless a problem occurs, such as when failing to hand tickets over to customers on time or when circulating forged tickets.

The tier-one agent authenticates the tickets that are printed and sold (or sold via its agents), before audiences enter the event venue. Prior to 2004, ticketing practices were essentially transactional – audiences' payments were collected by ticketing companies and their affiliated agents on behalf of the event organiser as a means of gaining permission to attend an event. Up until 2004, state-owned enterprises had been the key players in organising ticket distribution. However, after 2004, their market share has gradually been eroded by newcomers such as ChinaTicketCo. The practice of ticketing had, up to this point, been extremely stable and very little innovation or change had been introduced. One of the turning points was marked by ChinaTicketCo's introduction of online ticket sales for Faye Wong's 2004 concert in Beijing, which we describe next.

Phase two: Ticketing as a transactional and a relational practice

Faye Wong is a well-known Chinese, female performing artist, so the success of her ticket sales was predictable. This was the largest event that ChinaTicketCo had ever been assigned as the tier-one ticketing agent, as it had previously been a tier-two agent. The event could, therefore, make or break its reputation. There were several changes and challenges associated with the move from a tier-two to a tier-one ticketing agent. First, a tier-one agent was responsible for producing tickets that were distributed and sold to tier-two agents and customers. Substantial investment in ticket production facilities was required, in particular for producing tickets that were unlikely to be forged. Second, distribution costs became a major concern as a tier-one agent had to deal directly with a greater number of agents and customers than would be the case for a typical tier-two agent. Third, a tier-one agent was responsible for authenticating tickets when audiences entered the venue and it was crucial to develop a system capable of doing this accurately, reliably and efficiently. To try to win this contract, the founder of ChinaTicketCo had decided to introduce online sales and this is essentially what led to the company's successful bid. The introduction of online sales became a headline on its own, as it was the first ever attempt to do this in China (Figure 4.1).

Figure 4.1

New ways to sell, through online platforms

The online channel was exceptionally well received. To purchase tickets online, a customer was required to register and open an account with ChinaTicketCo. Compared with buying tickets via such traditional channels as the telephone or queuing at box offices, where, for a popular event, long wait times were common, the online sales system allowed ChinaTicketCo to smoothly handle a large number of transactions as customers clamoured to get tickets during the first few hours of ticket release. Moreover, the introduction of online sales was important not only in terms of its benefits for ticket sales but also because it subsequently afforded other innovations. For example, it provided a crucial mechanism for ChinaTicketCo to systematically develop, understand and maintain relationships with its customers, which it could then leverage for ticket sales of future events. Further, the following year, a large three-month event called Global Festival, targeting school children for the summer holiday period of 2005, sold more than 4 million tickets via the online channel.

The data collected through the online system enabled ChinaTicketCo to understand the market, forecast ticket sales and reach targeted customers more effectively than its competitors. This enabled ChinaTicketCo to be in a position to predict ticket sales for upcoming events better than competitors – a very important capability from the perspective of event organisers, thereby helping it to build strong relationships with event organisers and venues.

Phase three: Ticketing as a transactional, relational and experiential practice

The year 2009 marked another key milestone when several strategic initiatives were launched by ChinaTicketCo. These included a B2B (business-to-business) ticketing platform, e-tickets, online seat reservations and mobile check-in. The B2B ticketing platform was developed to streamline the transactional aspect of ticketing. Through this platform, the practice of issuing tickets changed since tickets could now be created, distributed and processed electronically with a minimal level of human intervention. By installing software and printers for each agent, the agents were able to issue tickets in situ. This new practice not only significantly reduced delivery costs but also helped to enhance customer satisfaction. The second change that was afforded by the adoption of a B2B platform was that ChinaTicketCo ventured into new areas of practice that included issuing tickets for different types of events and services, such as spa trips, ski passes, city breaks and tourist attractions. The rationale behind this new practice was that the basic elements of ticketing are similar across a wide range of products and services.

In addition to this improved ticket sales capacity, ChinaTicketCo started to use social media to connect with its customers. Most of the other ticket agencies did not feel that it was worth investing significant resources in social media, because most events would take place just once. However, this is not how ChinaTicketCo approached things. Instead, while recognising that strategically attempting to influence fan–star interactions through social media is extremely resource demanding, ChinaTicketCo nevertheless made this investment, believing that it could yield commercial gains. The company employed a team of 12 full-time and 55 part-time staff in a social media division. This team was responsible for over 1,300 accounts for different performing artists and celebrities on the Sina micro-blog site (similar to Twitter). The focus of the practice of these employees was to participate in various fan group blogging sites, posting messages and replies and generally monitoring activity. In particular, for an upcoming event, they would become active in that musician's site, encouraging fans to get excited about the event. They did this even for celebrities whose events they were not organising, in the hope that the knowledge gained through such activity might help them to organise events in the future, especially were the celebrity's fan base to increase in size as a result.

In addition to building and sustaining relationships with fans, another benefit enabled by social media was to nurture and consolidate the fans into a community. By encouraging fans to share their event photos, to gossip and to provide feedback and by helping to organise fans so that they could purchase tickets together to get discounts, ChinaTicketCo was able to enhance the experiences of its customers at the same time as it strengthened its relationships with them. Clearly, the transactional aspect of ticketing practice remains essential for commercial purposes. However, what we observe is how relational and experiential aspects had been simultaneously incorporated to create a complex bundle of new ticketing practices that align with an evolving strategy. We use this case in this chapter to guide the introduction of key topics for discussion.

Introduction: Strategy versus strategising

As with the case of ChinaTicketCo, organisations engage in digital innovation in order to become or remain competitive and so what and how they innovate depends on their strategy. Strategy has often been conceptualised as a grand vision, which is formally planned by the top executive team and then executed by others lower down in the organisational hierarchy (Prahalad and Hamel, 1994). Perhaps the best-known variant of this approach to strategy is presented by Michael Porter (Porter and Millar, 1985; Porter, 1991). Porter discusses how to identify an attractive market niche (based on his 'five forces' model) and then what business strategy to adopt to be successful within this niche – both choices (what niche and how to be successful in that niche) are based on pre-action planning. Innovation from this perspective is related to both choosing the market niche and selecting the particular strategy (and associated business model) to adopt within this chosen niche. We present these ideas in this chapter and contrast this view of strategy as a *deliberate* planning process (strategy *formulation*) with Mintzberg's

(Mintzberg and Waters, 1990) view of strategy as an *emergent* process (strategy *formation*). We also discuss the more recent *strategy-as-practice* view that sees strategy as something that is constantly unfolding in the flow of practices as undertaken by strategy practitioners (Whittington, 1996; Jarzabkowski, 2005).

Strategy as a deliberate planning process

In Porter's five forces model (Porter, 1991, 2008), he outlines the factors that determine the attractiveness of a market, with attractiveness determining the profits that can be generated in a given market niche (Figure 4.2).

A firm must therefore decide its strategy, and so in which niche to operate, based on an analysis of five forces:

1. *The threat of established rivals*

Industries differ in terms of the concentration of market share – in some industries only a few firms dominate, whereas in others lots of firms each hold a degree of market share. In the latter case, the market is described as fragmented and there is more competition (or *rivalry*) between firms. Other features of an industry also influence the degree of rivalry. For example, where market growth is slow, rivalry will be more intense because firms are fighting over the same customers rather than an enlarging customer base. Similarly, where customers can switch easily, rivalry will be high because firms can be more effective in persuading people to change. Rivalry will also be high when the exit barriers are high, meaning that a firm has so much sunk cost in people and equipment for producing products and services for a particular market that it wants to stay in this market at all costs (recall the example of Kodak in Chapter 3). However, since it is illegal for firms to collude (at least, in most Western countries), there will always be some rivalry between firms in any industry and so firms try to make competitive moves to

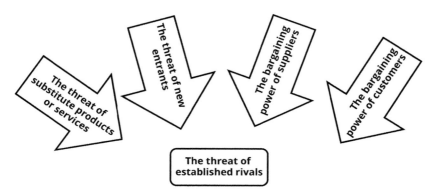

Figure 4.2

Porter's five forces

gain advantage over their rivals. Innovation, and so knowledge exploration and exploitation, is key to these competitive moves, whether the innovation relates to a product or a service (including pricing, promotion and distribution) or to the processes of production and organisation.

2. *The threat of substitute products or services*

Rivalry occurs not only between firms within the same industry but also with firms in other industries if they produce substitute products or services (e.g., drinks can be packaged in aluminium, glass, steal, plastic or even paper, so there is rivalry between firms in these different industries). Such threats of *substitution* generally lead to price competition but can also cause industries to innovate so that they are preferred on another criterion. For instance, industries can compete on their environmental sustainability (e.g., plastic bags that are biodegradable can compete with paper bags that can be recycled) or their ability to protect or preserve a product (e.g., polystyrene keeps a hot drink hotter for longer and so may be preferred over paper even though paper is more environmentally friendly) or, as with ChinaTicketCo, their ability to distribute their product or service more easily using digital innovations. Again, attempts to compete against potential substitute products or services can therefore stimulate digital innovation.

3. *The threat of new entrants*

While there is threat from firms within an industry and firms in substitute industries, there is also a threat of new firms entering an industry and taking market share. How significant the threat of *new entrants* is depends on barriers to entry within a given industry. Some barriers relate to physical resources. For instance, if a lot of expensive and industry-specific equipment is needed to enter a market, this creates a barrier to entry. A specific example here would be aerospace – it takes huge resources to invest in space travel and firms already active in the industry therefore have significant advantage over potential new entrants. There are also knowledge-related barriers (e.g., where an industry is based on patents and other types of proprietary knowledge, newcomers face barriers to entry). The examples we introduce in Chapter 5 regarding drug development illustrate this issue when the importance of owning and protecting Intellectual Property (IP) is discussed. However, pharmaceuticals is also an interesting industry that has problems with the threat of new entrants since once patent protection on a product has ended, new entrants can come in and make generic versions of the branded drug. These generic products are sold at a much-reduced price because the generic drug manufacturer does not have to bear the costs of all the research and development work, including the very expensive clinical trials that are required by regulation.

Besides patents, a more in-depth knowledge 'gap' might exist between those firms that have been operating in a specific market/industry for years and new entrants. The latter might not have the knowledge ('core competencies' and know-how) that is necessary to succeed or even simply to survive. As another example of the threat of new entrants, the airline industry is currently being disrupted by the introduction of the new Dreamliner aeroplane, which makes it more economical to fly longer distances and so is opening up competition on intercontinental routes for the cheaper airlines to enter this industry segment. The actions of a budget airline, Norwegian, are an interesting example of this. Norwegian has moved into transatlantic routes, undercutting the fares charged by established airlines such as American Airlines, British Airways and Virgin. In this instance, the threat of new entrants has increased because of innovation by the airline manufacturers.

4. *The bargaining power of suppliers*

Industries differ in terms of the power of suppliers – whether they are suppliers of raw materials, components or labour. For example, if an industry requires highly specialised labour that is in relatively short supply, employees can demand higher wages and so reduce the profits that can be accrued. Or, if suppliers are highly concentrated, a vendor firm will have less ability to 'shop around' to reduce dependence on a particular supplier. Therefore, there are some industries where there is heavy dependence on a single supplier. An example is the agricultural seed production industry, which is dominated by a small number of firms, such as Monsanto. Monsanto has developed (and patented) seed varieties that are based on genetic engineering to try to create crops that, for example, can withstand drought or produce greater yields. There is considerable controversy over this type of innovation because, for instance, we don't know the long-term effects of consuming genetically modified (GM) crops on human health, nor do we know the impact of growing such crops on the natural environment. At the same time, there are some clear benefits of growing such crops (e.g., allowing crops to grow where otherwise it would not be possible). This example illustrates that innovations can also be highly controversial.

The documentary *Seeds of Death*, about Monsanto, illustrates the dark side of innovation. More importantly, from the point of view of the *bargaining power of suppliers*, this example demonstrates how a dominant supplier makes it more difficult for those dependent on the supplier to opt out of the innovations that are presented by that dominant supplier or to innovate themselves. For example, in the USA, Monsanto licenses its GM corn and soybean seeds so that 80 per cent of US corn and over 90 per cent of US soybean are attributable to Monsanto. This monopolisation makes it difficult for farmers to grow other varieties and act independently of Monsanto – they are subsequently at the mercy of any Monsanto price increases. Monsanto has achieved this dominance by taking an aggressive stance on patenting its seeds and then enforcing its IP by suing farmers in certain circumstances, for example where they plant alternative seeds that include some of the Monsanto patent-protected seed.

5. *The bargaining power of customers*

In some industries, it is not the suppliers that have the power but the customers. The *bargaining power of customers* occurs in industries where there are only a few customers (or even just one customer – such as in the defence industry, reliant as it is on government contracts) or only a few very powerful customers. In retail, for instance, some of the very large supermarket chains have considerable power because suppliers need their products in the large chain stores. Similar forces are at play in the automotive industry (see, for example, Webster (1995)), where major car manufacturers have considerable power over their suppliers (for component parts etc.). If buyer power is high, they can squeeze the price and this can tend to promote suppliers to focus on innovations that help to improve efficiency savings. The dairy industry in the UK is an example here, where dairy farmers have been 'squeezed' because a price war in the large supermarkets has helped to depress the price that farmers receive for their milk, even while their costs (largely for animal feed) have gone up. The result is that there has been a fall in the number of dairy farmers, with those remaining having to find ways to innovate in order to become more efficient (illustrating the process of natural selection as predicted in the theory of population ecology (Hannan and Freeman, 1977), see Chapter 3).

A firm, then, must decide which line(s) of business to be in, based on an analysis of these five forces. In other words, Porter's five forces model is helpful for a firm to understand how to be positioned in a particular industry with respect to competitors, buyers, suppliers, type of market and products/services developed. This, of course, is problematic in the sense that an industry is not static – any of these forces can change, as when there is an unpredictable event of some kind. For example, an economic downturn can be associated with a rapidly shrinking market and so competition increases, and some firms may collapse as a result. Another example can be related to a sudden change in the demand arising from factors unrelated to the economy: think of global airlines after the 9/11 attacks in the USA in 2001, which made this industry suddenly a very challenging place to succeed. The changes in this industry demanded lowering prices (to try to sell more tickets) but also involved expenses associated with being compliant with new laws and regulations (e.g., secure, bulletproof cockpit doors equipped with locks and new access policies). However, the airline industry is also very accessible in that barriers to new entrants are low. For instance, for a relatively low initial capital outlay, it is possible to rent a gate and an aircraft and to fly passengers on a single route connecting two major cities.

The above two examples are illustrative of how environmental changes might require firms to rapidly reconfigure their strategies. But, for the purpose of this chapter, a good example of issues that reflect the natural 'becoming' (i.e., industries and markets are always emergent and in a constant state of flux) of industries and markets is related to disruptive innovation, which may change any or all of these forces overnight. The Kodak example in Chapter 3 provides an illustration of how a company can be undone by a sudden change in the competitive landscape – in this case, the threat was from digital photography that undermined the film market that Kodak had thrived on for years. Nevertheless, the idea behind Porter's framework is that a firm should acquire knowledge about each of these forces and, based on this, make a decision to enter (or exit) a particular market niche. So, Kodak could have used this framework to more quickly understand the threat from digital photography as a substitute product. This is not a one-off decision, however: firms can change the market niche they are in or diversify so that they are in several markets simultaneously and can enter or exit based on an analysis of these five forces. The example of ChinaTicketCo moving to sell tickets for other types of activities, not just events, is an illustration of how firms can diversify.

Firms that don't do their homework in this respect can be caught out by one or several of the forces that then restrict their ability to make a profit – as was the case for Kodak. As another example, Webvan, an online 'credit and delivery' grocery business that went bankrupt in 2001, attempted to break into the US food retail market with an online home delivery service but failed because the investment it made in infrastructure was not matched by the size of the customer base it was able to attract. Buyers did not switch to the online service in the numbers anticipated because they preferred the flexibility of going by car to the supermarket to do a 'big shop' themselves rather than having to wait at home for the period of a scheduled delivery. Webvan's analysis of its market niche failed miserably. This can be contrasted with ChinaTicketCo, where the company's introduction of online ticket sales was much more effective.

The Webvan example refers to an incorrect estimate of the *switching costs* associated with a substitute service. Switching costs are the costs that keep a buyer tied to a firm, because it costs too much to change to an alternative that offers similar or equivalent services/products. For instance, having several phone accessories (home/car chargers, Bluetooth speakers/devices etc.) might lead a customer to stick with a particular phone brand because it is too expensive to buy the same accessories for another phone given that accessories are generally not compatible across brands. In the case of Webvan, customers were not willing to give up the flexibility of shopping in person, with many deciding to stick to their old habits. However, this failure has subsequently helped other companies, such as Amazon, to better approach new online retailing opportunities, as explained in a 2013 *Forbes* article entitled 'Four Lessons Amazon Learned From Webvan's Flop'. This example showcases the relevance of being able to *execute* and not simply *conceive* an innovative idea (Figure 4.3).

Porter's three strategies

Another of Michael Porter's notable contributions to strategy concerns his 'three strategies': *differentiation* or *cost leadership and segmentation*. Once a firm has made a decision about which industry or industry niche to compete in, it must then decide *how* to compete. Porter acknowledges that selecting an attractive market niche is only part of the strategy story because firms must also develop the competencies and an appropriate business model to

Figure 4.3

Conception and execution of a strategy
©Blend

exploit the potential of this niche market. He identifies different approaches that a firm can take to achieving competitive advantage within its chosen line of business. Essentially, there are two aspects to this. Firms must decide whether:

1. To focus on being lower-cost than rivals or differentiating products and services so that they have distinctive features for which customers are willing to pay extra. Walmart is the classic example used to illustrate a *low cost strategy*, whereas an example of a *differentiation strategy* would be Wholefoods, a US food retailer, which in 2017 merged with Amazon. Wholefoods sells natural and organic foods and is committed to sustainable products. ChinaTicketCo's strategic moves also suggest that it has opted for a differentiation strategy, adding the relational and experiential aspects of buying tickets to the original transactional aspect.

2. To adopt a more or less focused strategy (concentrating on a small and specific niche or a mass audience). As an example, in higher education, some colleges focus on specific niches – such as business or liberal arts – whereas others attempt to cover all subject areas and are comprehensive in nature. ChinaTicketCo has moved from a more focused strategy – based on event ticket sales – to a broader strategy – selling tickets for all sorts of events and activities.

A firm, then, gains competitive advantage to the extent that it can use resources and competencies to perform at a higher level than others within the same chosen line of business or market niche and adopting a business strategy. If the selected strategy is low-cost, the firm must develop processes that ensure a lower production cost that will reflect either a lower price to customers or a higher margin – either way, the result is that the firm will be more efficient than its competitors. A differentiation strategy, on the other hand, involves offering products, services or features of products or services, which are more distinctive than those that rivals offer. Dell is often used as a classic example of this strategy, with its mass customisation approach, allowing it to build computers to match the specific needs of customers.

A differentiation strategy relies on an innovation approach that allows the firm to create differentiated products or services and can be based on either incremental improvements in customer offerings or more radical or even disruptive offerings that undermine others in the market (Christensen and Overdorf, 2000). Of course, it is important to note that a low-cost strategy also requires innovation because to reduce costs below one's competitors typically means organisations needing to introduce some kind of change. Walmart, for instance, has been innovative in the ways it manages its supply chain to reduce costs through data sharing with its suppliers, so that they can better forecast demand and so have supply ready when and where it is needed, and improved logistics using, for example, a technique called *cross-docking*. Cross-docking involves having goods delivered to stores without them sitting for long periods in warehouses. Both these innovations help Walmart to reduce its own costs, which it can then pass on to the consumer. This underpins the idea introduced previously that today firms often need to be *ambidextrous* to survive in highly competitive, global markets (see Chapters 1 and 3) (Figure 4.4).

The approach to strategy that underpins Porter's work, then, is based on the idea that strategy is a *deliberate* process that involves careful planning of both the niche to work in and the business model to be used to gain competitive advantage in this niche. The assumption is that careful planning of these fundamental aspects of strategy is what matters. Of course, in terms of the selected business model, developing the competencies that will enable the business model to be enacted are seen to be important, but this is still treated as a planning process that involves identifying what these core competencies are and then planning to ensure that the firm 'has' these competencies (e.g., by hiring people with the right skills and experience or by investing in the appropriate technologies). Knowledge in this sense precedes strategy, as strategy is based on having 'good' knowledge – on doing one's homework so that strategic plans can support the prosperity of the firm.

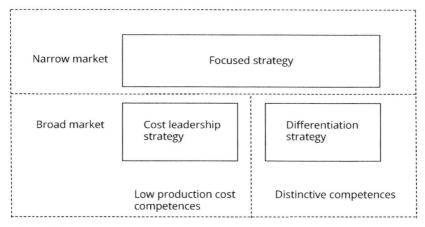

Figure 4.4

Porter's three generic strategies

 Key Concepts: Planned Strategy

Porter's five forces:

- Rivalry – how much rivalry there is between incumbent firms in the market niche
- Substitution – how easy it is to substitute for a product or service in the market niche
- New entrants – how easy it is for new entrants to penetrate the market niche
- Bargaining power of suppliers – how much power key suppliers have in the market niche
- Bargaining power of customers – how much power customers have in the market niche

Porter's three strategies:

- Cost leadership – competing on the basis of lowest costs of the product or service
- Differentiation – competing on the basis of some unique aspect of the product or service offered (even if the cost is higher)
- Segmentation – focusing on a very small segment of the market or a broader segment

An alternative, emergent view of strategy

For quite some time, however, there has been another approach to strategy at work. This sees strategy as something that is more *emergent* than the above deliberate, planned approach would suggest. Henry Mintzberg recognised that it is really hard, in practice, to plan a firm's strategy up-front given changing circumstances, and so he stressed the need to recognise (both practically and theoretically) that a strategy *emerges* over time as 'a pattern in a stream of decisions' (Mintzberg and Waters, 1985, p. 257). This allows us to differentiate between *intended* (what the plans say) and *realised* (what is actually achieved) strategies – the two often being quite different in nature. This is because some parts of a strategy may not be realised in actuality (e.g., because a competitor brings out a product that leapfrogs your product) or because some plans emerge from experience (e.g., when a manager 'tweaks' some aspect of the manufacturing process and identifies the opportunity to add new features to a product that were not in the original plans). This emergent view of strategy, in other words, recognises that learning based on the execution of some strategic plan results in the creation of knowledge that can produce new insights that were not available prior to this actual execution. This demonstrates the way in which knowledge-as-a-possession is a tool of knowing (see Chapter 1), with the knowledge in this instance related to how to obtain competitive advantage for a firm. You can see this emergent nature of strategy in the ChinaTicketCo case – the company would not even have been able to think about venturing into new types of ticket sales before it had moved to online ticket sales in the first place.

The more recent manifestation of this emergent view of strategy has taken this thinking a step further, by considering *strategy-as-practice* (Whittington, 1996; Jarzabkowski, 2005). The strategy-as-practice approach emphasises the day-to-day activities of practitioners who shape, refine and actualise strategy through what they *do*. A strategy (or better, *strategising*) is viewed as an emergent set of practices, which are constantly in the making (Jarzabkowski, 2005; Jarzabkowski and Paul Spee, 2009). Whittington (2006) outlines the need to examine three conceptual elements and their interactions in adopting a strategy-as-practice perspective: *practitioners* (the people who do the strategy work); *practices* (the routines – social, symbolic and material – that guide the strategy work); and *praxis* (the flows of actual activity through which strategy is achieved).

Strategy *practitioners* include actors who are directly engaged in the shaping and enacting of strategy, as well as individuals, and other institutions, such as policy-makers and regulatory bodies, that have direct or indirect influence on what might be feasible and legitimate (Jarzabkowski and Whittington, 2008). Practices are 'embodied, materially mediated arrays of human activity centrally organised around shared practical understanding' (Schatzki, 2001, p. 2). Practices can best be understood as *routines* – 'scripts' (not necessarily written) that allow individuals to avoid constant learning in daily work activities, because routines are based on prior learning and experience. These routines provide a guide to what should be done in a particular context based on existing cultural rules, languages and procedures, supported by material objects, in particular in today's work environment and information technologies. In Chapter 5, we discuss this idea of routines in more detail and introduce the idea of contexts being 'equipped' to support the routines of work.

When we discuss routines, it is essential to understand that a routine or 'script' provides guidance on how to undertake a certain organisational activity (e.g., a practice such as negotiating with a customer); however, all salespeople have their own style of conducting a negotiation to close a contract. While the routine is the *practice*, the way the routine is enacted is the *praxis*. The distinction between practice and praxis is essentially the difference between the routine that guides activity and the actual activity itself (Reckwitz, 2002): the script and the actual play, to use a theatrical analogy. It is necessary to distinguish between practice and praxis because the praxis – the actual activity at any point in time – may be more or less similar to the practice. This is because strategy praxis is the play in the theatre, which provides room for interpretation (even when actors are following a literal script), and because there may be a need to improvise due to unanticipated circumstances (e.g., because a prop is mishandled or a line is forgotten). Praxis therefore depends on the unique interplay of practices and practitioners in each rendition of the play. Most importantly, praxis accounts for the fact that while organisational activities may be institutionalised (DiMaggio and Powell, 1983; Powell and DiMaggio, 1991, 2012), routines nevertheless can transform quite dramatically over time (Feldman and Pentland, 2003) – as we discuss next when considering *strategy-as-practice* and *ambidexterity*.

Strategy-as-practice and ambidexterity

Since we have already noted the importance of ambidexterity for most organisations today, we can consider how the strategy-as-practice perspective would help us understand how this can be achieved. Given the elements of the practice view outlined in this chapter, it is important that we think about all three elements when linking strategy-as-practice to ambidexterity: (1) the strategy *practitioners* who are engaged in the shaping of ambidexterity; (2) the *practices* of those involved that allow practitioners to simultaneously explore and exploit knowledge; (3) *praxis* – the practice enacted in a moment in time. Since the essence of ambidexterity can be seen as the concurrent enactment of knowledge exploration and exploitation to achieve complementary effects (Cao et al., 2009), we need to look not at single practices but rather examine the ways practices are bundled and re-bundled together over time (Jarzabkowski and Spee, 2009). In terms of this last point, Nicolini's (2011) use of the idea of a particular 'site' of knowing can be helpful.

A *site of knowing* is a social and relational space where various practices come together as practitioners with different but complementary interests coordinate their efforts in order to achieve particular objectives. Of course, this 'coming together' may be more or less successful, but unless someone is trying to sabotage things, in a particular site of knowing those involved attempt to work together to produce the intended objective. Nicolini (2011), for example, examines a site of knowing in a hospital where the different practices of various professional groups (e.g., nurses and doctors) are played out (i.e., the praxis) to achieve (mostly) successful patient-care practice. If one were only to look at the practices of doctors rather than the practices of doctors in conjunction with the practices of other practitioners in a particular site of knowing, one would attain only a partial understanding of practice. The same can be said for the ChinaTicketCo case – if we looked only at the practices of employees, we would miss how the company's interactions with fans on social media have influenced its ability to develop new ways of building customer experiences that can enhance the value it provides to event organisers.

Figure 4.5 is an example we use in this chapter to demonstrate practice. A doctor initially follows a routine (practice) and has in mind a specific medication for a patient; however, the doctor then interacts with a nurse and a psychologist (other practitioners in the site of knowing) and learns other important details about their patient. The doctor examines the patient further using digital technology (a scan, which is a material actor in this example; see Chapter 7). From these interactions, the doctor *learns* – through the knowledge exploration and exploitation practices occurring in this site of knowing. Finally, the doctor comes up with a diagnosis that is slightly different from the initial intent to prescribe a medicine (their top-down, planned strategy). Perhaps the dosage has changed in light of the psychological situation of the patient (knowledge exploitation, usage of existing knowledge acquired from a peer – another doctor). Or perhaps the nurse has advised the doctor that the patient was allergic to a component of the initial

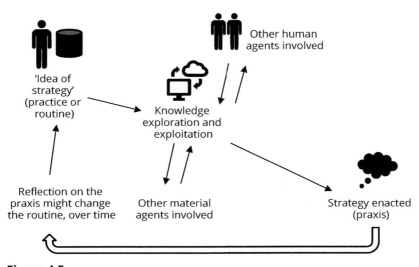

Figure 4.5

Practice, praxis and knowledge exploration and exploitation

prescription, so the doctor needed to talk to the hospital's lab so they could provide a medication with the same active ingredient but without the component that would create problems for the patient (knowledge exploration, discovering a way, or 'workaround' to a problem). All these practices have followed a script (the basic steps that a doctor needs to take before signing off a prescription). However, these practices led to a particular outcome (the final prescription) that was different from what the doctor was initially going to do. The praxis contributes to the doctor's knowledge and might change their way of approaching a problem in the future – a change of routines, in other words (Feldman and Pentland, 2003). This example is also helpful to understand the emergent nature of knowing, through practices (enactment of routines) that generate (emergent) praxes. In relation to the day-to-day running of a ward, one can look at how practitioners coordinate their activities and so exploit existing knowledge about patient care. One can also then look at how knowledge is explored to create innovations in this ongoing site of knowing.

The following example is slightly different and therefore offers another useful way of thinking about this topic. In a paediatric ward, there was an ongoing practice of handover between two shifts – all clinical staff were supposed to come together in a room by the side of the ward and the outgoing shift clinicians (the practitioners) discussed each patient in turn with the incoming shift clinicians in order to ensure continuity of care. Actual practice, however, was often disrupted – those finishing their shift were still trying to complete their tasks (such as the filling out of a prescription request) while those coming in were more or less attentive, again depending on other things they were trying to do simultaneously (like gather the equipment they would need to start their ward round). In practice, then, everyone's attention was not entirely on the handover, which could take up to 30 minutes. This had led, on several occasions, to a patient's needs not being clearly communicated. One of the doctors had learnt of a traffic light system used in a different context to categorise patient illness severity. She decided to translate this idea to her site of knowing and devised a traffic light system where patients were categorised as green (little information needed to be handed over), amber (some but limited information was needed) and red (considerable attention was needed, with particular focus being given to these patients). In this simple approach, attention at handover could be concentrated where it was most needed (i.e., on the 'red' patients), and the problems of coordination between the two shifts were greatly reduced. In this example, we see how it is that the coordination of practices, or lack thereof, between different practitioners is of considerable importance. This helps us to understand ongoing routines of practice but also the generation of new practices – the exploitation *and* the exploration of knowledge, in other words. This traffic light system is now a common practice to help people differentiate the importance of different practices; often, these traffic light systems are digitised on dashboards that identify current performance against agreed benchmark indicators. The introduction of such traffic light systems shows that a site of knowing is not fixed, so that, given the emergent nature of praxis, a site can *shift* over time (Nicolini, 2011) in the sense that new practices and practitioners and new relationships between practices and practitioners are introduced. When a site shifts, some of the taken-for-granted assumptions and norms of practice can become invalid and create a space for debate and contestation among its practitioners (Contu and Willmott, 2003); as a result, site shifting potentially allows the emergence of new capabilities. In our traffic light example, for instance, the nurses took over responsibility for categorising the patients as red, amber or green and developed new capabilities as they did so. Again, when we think about ChinaTicketCo, we can see how practices and practitioners are bundled and re-bundled over time to allow the company to explore new opportunities (e.g., the bundle of practices related to online booking opened up opportunities for targeted marketing given that the company now had a database of customers, with information on their previous ticket purchases and preferences). In this case, the focus is on digital strategising because ChinaTicketCo uses IT to both *exploit* existing products and services and *explore* new opportunities for changed products and services. We turn to this issue of digital strategy and strategising next.

 Key Concepts: Strategy Formation

Mintzberg:

– Emergence – strategies emerge over time in response to events that may not have been anticipated

– Realised versus intended strategies – realised strategies (the strategies that are actually enacted) may be different from the intended (i.e., the planned) strategies

Strategy-as-practice:

– The *doings* of strategy – strategy unfolds through practice as an outcome of interactions between practitioners, practices and praxis

Digital technology and strategy-as-practice

It is interesting to focus specifically on IT strategy and strategising because, as we have stressed, so many products and services include some digital technology today, even those products not conventionally considered to be digital, such as a car. Although some, notoriously, have argued that 'IT doesn't matter' (Carr, 2003), believing it to be little more than a commodity (like electricity – widely available to all), Bill Gates was alleged to have called Carr's article 'the dumbest thing I ever read'. Why this difference of opinion? Well, from the perspective of the commoditisation of digital technologies there are few distinguishing features of such systems when adopted across different organisations. For example, applications, such as an enterprise system like SAP, can be bought and implemented 'off-the-shelf'. However, from the perspective of strategy-as-practice, it is what we *do* with the technology – how we manage our digital services, how we build our digital capabilities (Peppard and Ward, 2004) and how agile we are in developing and using digital technologies (Desouza, 2006) – that matters and that can set us apart from the crowd (Galliers and Leidner, 2009). Therefore, some organisations are perceivably better than others in utilising business intelligence systems to good effect by delving into the data and questioning taken-for-granted assumptions about the decisions and actions being taken within the business (Shollo and Galliers, 2016). ChinaTicketCo provides a good example of using digital technology to continuously 'reinvent', in this case, the business of ticketing. In Chapter 9, we consider the rise of sensor technologies that are now incorporated in so many products and services.

Recent articles on Information Systems (IS) strategy-as-practice have been published as a consequence of this kind of thinking (Whittington, 2003; Galliers, 2011; Peppard et al., 2014; Marabelli and Galliers, 2017; Morton et al., 2018). This strategy-as-practice view of IS questions more traditional approaches to strategy that focus on exploiting digital technology for competitive advantage based on a deliberate planning process (e.g., Porter and Millar, 1985). The more traditional view of strategy tends to focus on the *exploitation* of digital technology in line with business imperatives and is 'top-down' in nature in that the digital strategy attempts to align itself with predefined business strategies (Henderson and Venkatraman, 1993). You can probably already see that this might be difficult given our discussion in this chapter about how strategies constantly emerge from ongoing practice.

Understanding digital strategising as an ongoing, dynamic process of strategy – thinking and doing – takes account of organisational as well as environmental changing circumstances (e.g., innovations such as new products, services and processes arising from the astute use of new digital technologies, as in the ChinaTicketCo case). Given the acknowledgement of emergence, this strategy-as-practice view also recognises that there will sometimes be unexpected consequences – good and bad, but in neither case what was planned – from the adoption and usage of new digital technologies and systems. Robey and Boudreau (1999), for example, talk about the unexpected consequences of the introduction of digital systems.

A good example of this emergence can be seen in the use of emails to save documents, a function that was an unexpected consequence of the introduction of email. So, if a document is important for me, I – like many others – send it to myself so that it is available from my mail server, even were I to lose my computer or if my computer crashes. Yet, from a strategic perspective, email was not introduced to serve this purpose. This example, along with being illustrative of how a digital technology can be used in ways for which it was not designed (e.g., email was designed to send/receive messages, not for backup purposes), sheds light on how firms should be prepared to reconfigure their strategy because of how employees, customers or other stakeholders *enact* digital technology. To this end, the intense use of emails to back up files led Google to provide users with a virtually unlimited email account. It was initially just 1 GB – hence the name, Gmail. Once internet infrastructure became a reliable means by which to exchange files quickly, and 24/7, IT companies 'invented' the Cloud, as with the example of Dropbox in Chapter 2. In this instance, the innovation was related to a business model rather than a new technology since this technology of online file sharing had been around for quite some time (Figure 4.6).

From the above, we learn that, from the strategy-as-practice perspective, digital technology cannot align with some fixed business strategy, but rather the focus is on *aligning* (Karpovsky and Galliers, 2015), which involves a continuous back-and-forth between the business strategy and emergent opportunities from the *exploration* as well as exploitation of digital technologies – and this can happen at almost any level of the organisational hierarchy. The case of backing up files using mail servers, and the consequent introduction of the Cloud, highlights the twofold nature of digital strategising in practice (with knowledge exploration alongside exploitation). You can find other examples of this emergent strategising in the ChinaTicketCo case.

Figure 4.6

New technologies such as Cloud computing enable emerging innovation
© *Getty Images/iStockphoto*

The key concepts of the strategy-as-practice idea as applied to digital are provided below and can be useful in examining the questions about the ChinaTicketCo case that relate to this view of digital strategising.

 Key Concepts: Strategy-as-Practice

Digital Strategy Practices:
Institutionalised routines that guide digital strategic activity, based on traditions, norms and procedures that exist both within the organisation and beyond its boundaries

Strategy Practitioners:
Those individual actors who shape and actualise digital strategy, including actors within a focal organisation but also, for example, external policy-makers, regulatory bodies and competitor organisations

Digital Strategy Praxis:
The actual activity of creating and enacting a digital strategy that may be more or less similar to the institutionalised routines because of the sense-making/interpretation of the particular practitioners involved and because of unanticipated events that can disrupt routine practices

Strategy Site:
The social and relational space where digitally enabled practices are bundled together in particular ways by the practitioners involved and that can change over time as an outcome of praxis

Conclusions

In this chapter, we have seen how the strategy-as-practice perspective is very different from the strategy-as-planning perspective, particularly in relation to its understanding of how knowledge and knowing interact. Whereas the latter sees knowledge as something that precedes the planning, essentially *feeding* the plan, the former treats strategising as something *emerging* from knowing in practice. As with the mutual constitution view of knowledge itself (knowledge as a tool of knowing), we can probably see that both planning and emergence are relevant to effective strategy. While organisations need to engage in deliberate planning in relation to their digital innovation ambitions and seek and use knowledge about their industry and business models in so doing, they also need to recognise that no amount of careful planning will ever be sufficient for effective strategising. This is so because *knowing in practice* will produce new insights that may warrant modification or even abandonment of the plans for innovation. Indeed, sticking to the plan may well be a recipe for disaster, as with the example of Kodak that we discussed previously.

Case questions

The following case questions might also be relevant to using this chapter in teaching exercises and discussions:

1. Apply Porter's five forces model and competitive advantage ideas to ChinaTicketCo. How much of the firm's success does this explain?

2. What might be the threats to ChinaTicketCo's position of pre-eminence in its niche market?

3. What does the case tell us about the way practitioners and practices shape and are shaped by strategy? Apply the elements from the key concepts box on strategy-as-practice to this case to examine strategising processes at ChinaTicketCo.

4. What role does IT have in the emergence of the strategy at ChinaTicketCo?

Additional suggested readings

Jarzabkowski, P. and Spee, A. 2009. Strategy-as-practice: a review and future directions for the field. International Journal of Management Reviews, 11, 1, 69–95.

Nicolini, D. 2011. Practice as the site of knowing: insights from the field of telemedicine. Organization Science, 22, 3, 602–620.

References

Cao, Q., Gedajlovic, E., and Zhang, H. 2009. "Unpacking Organizational Ambidexterity: Dimensions, Contingencies, and Synergistic Effects," *Organization Science* (20:4), pp. 781–796.

Carr, N. G. 2003. "IT Doesn't Matter," *Harvard Business Review* (81:5), pp. 41–49.

Christensen, C. M., and Overdorf, M. 2000. "Meeting the Challenge of Disruptive Change," *Harvard Business Review* (78:2), pp. 66–77.

Cohan, P. 2013. *Four Lessons Amazon Learned from Webvan's Flop*. Forbes, Available at: https://www.forbes.com/sites/petercohan/2013/06/17/four-lessons-amazon-learned-from-webvans-flop/#2bb0f4748147.

Contu, A., and Willmott, H. 2003. "Re-Embedding Situatedness: The Importance of Power Relations in Learning Theory," *Organization Science* (14:3), pp. 283–296.

Desouza, K. 2006. *Agile Information Systems*. Oxford, UK: Elsevier.

DiMaggio, P. J., and Powell, W. W. 1983. "The Iron Cage Revisited: Institutional Isomorphism and Collective Rationality in Organizational Fields," *American Sociological Review* (48:2), pp. 147–160.

Feldman, M. S., and Pentland, B. T. 2003. "Reconceptualizing Organizational Routines as a Source of Flexibility and Change," *Administrative Science Quarterly* (48:1), pp. 94–118.

Galliers, R. D. 2011. "Further Developments in Information Systems Strategizing: Unpacking the Concept," in *The Oxford Handbook of Information Systems: Critical Perspectives and New Directions,* R.D. Galliers and W.L. Currie (eds.). Oxford: Oxford University Press, pp. 329–345.

Galliers, R. D., and Leidner, D. E. (eds) 2009. *Strategic Information Management: Challenges and Strategies in Managing Information Systems*. 4th edition. London: Routledge.

Hannan, M. T., and Freeman, J. 1977. "The Population Ecology of Organizations," *American Journal of Sociology* (82:5), pp. 929–964.

Henderson, J. C., and Venkatraman, H. 1993. "Strategic Alignment: Leveraging Information Technology for Transforming Organizations," *IBM Systems Journal* (32:1), pp. 472–484.

Huang, J., Newell, S., Huang, J., and Pan, S. 2014. "Site-shifting as a source of ambidexterity: Empirical insights from the field of ticketing," *The Journal of Strategic Information Systems* (23:1), pp.29–44.

Jarzabkowski, P. 2005. *Strategy as Practice: An Activity Based Approach*. London: Sage.

Jarzabkowski, P., and Paul Spee, A. 2009. "Strategy-as-Practice: A Review and Future Directions for the Field," *International Journal of Management Reviews* (11:1), pp. 69–95.

Jarzabkowski, P., and Whittington, R. 2008. "A Strategy-as-Practice Approach to Strategy Research and Education," *Journal of Management Inquiry* (17:4), pp. 282–286.

Karpovsky, A., and Galliers, R. D. 2015. "Aligning in Practice: From Current Cases to a New Agenda," *Journal of Information Technology* (30:2), pp. 136–160.

Marabelli, M., and Galliers, R. D. 2017. "A Reflection on Information Systems Strategizing: The Role of Power and Everyday Practices," *Information Systems Journal* (27:3), pp.347–366.

Mintzberg, H., and Waters, J. 1990. "Studying Deciding: An Exchange of Views between Mintzberg and Waters, Pettigrew, and Butler," *Organization Studies* (11:1), pp. 1–6.

Mintzberg, H., and Waters, J. A. 1985. "Of Strategies, Deliberate and Emergent," *Strategic Management Journal* (6:3), pp. 257–272.

Morton, J., Stacey, P., and Mohn, M. 2018. "Building and Maintaining Strategic Agility: An Agenda and Framework for Executive IT-Leaders," *California Management Review* (61:1). In Press.

Nicolini, D. 2011. "Practice as the Site of Knowing: Insights from the Field of Telemedicine," *Organization Science* (22:3), pp. 602–620.

Peppard, J., Galliers, R. D., and Thorogood, A. 2014. "Information Systems Strategy as Practice: Micro Strategy and Strategising for IS," *The Journal of Strategic Information Systems* (23:1), pp. 1–10.

Peppard, J., and Ward, J. 2004. "Beyond Strategic Information Systems: Towards an Is Capability," *The Journal of Strategic Information Systems* (13:2), pp. 167–194.

Porter, M. E. 1991. "Towards a Dynamic Theory of Strategy," *Strategic Management Journal* (12:S2), pp. 95–117.

Porter, M. E. 2008. "The Five Competitive Forces That Shape Strategy," *Harvard Business Review* (86:1), pp. 25–40.

Porter, M. E., and Millar, V. E. 1985. "How Information Gives You Competitive Advantage," *Harvard Business Review* (July–August), pp. 149–152.

Powell, W. W., and DiMaggio, P. J. 1991. *The New Institutionalism in Organizational Analysis*. University of Chicago Press.

Powell, W. W., and DiMaggio, P. J. 2012. *The New Institutionalism in Organizational Analysis*. University of Chicago Press.

Prahalad, C. K., and Hamel, G. 1994. "Strategy as a Field of Study: Why Search for a New Paradigm?," *Strategic Management Journal* (15:S2), pp. 5–16.

Reckwitz, A. 2002. "Toward a Theory of Social Practices: A Development in Culturalist Theorizing," *European Journal of Social Theory* (5:2), pp. 243–263.

Robey, D., and Boudreau, M.-C. 1999. "Accounting for the Contradictory Organizational Consequences of Information Technology: Theoretical Directions and Methodological Implications," *Information Systems Research* (10:2), pp. 167–185.

Schatzki, T. R. 2001. *The Practice Theory*. London: Routledge.

Shollo, A., and Galliers, R. D. 2016. "Towards an Understanding of the Role of Business Intelligence Systems in Organisational Knowing," *Information Systems Journal* (26:4), pp. 339–367

Webster, J. 1995. "Networks of Collaboration or Conflict? Electronic Data Interchange and Power in the Supply Chain," *The Journal of Strategic Information Systems* (4:1), pp. 31–42.

Whittington, R. 1996. "Strategy as Practice," *Long Range Planning* (29:5), pp. 731–735.

Whittington, R. 2003. "The Work of Strategizing and Organizing: For a Practice Perspective," *Strategic Organization* (1:1), pp. 117–126.

Whittington, R. 2006. "Completing the Practice Turn in Strategy Research," *Organization Studies* (27:5), pp. 613–634.

5 PROJECTS AND TEAMING

Summary

This chapter focuses on the prevalence of organising work through projects in today's organisations and some of the reasons for this. It also explicitly considers how innovation projects can be managed when there is uncertainty (including unknown-unknowns), where plans and controls can stifle rather than support innovation in the face of continuously emergent knowledge. The SkinTech case at the beginning of this chapter illustrates several relevant uncertainties. This is the only case in this book not covering digital innovation. We use this case because regenerative medicine is such a good example where there are many unknowns that influence the innovation process, as we illustrate. Other examples in this chapter provide examples of digital innovations that also experience similar unknowns that influence how the innovation unfolds. In considering these unknowns, we discuss the idea of learning plans rather than project plans, where people involved in innovation projects continuously experiment and adapt assumptions as more is understood over time about the potential for some new product, process or service. We will also explore the 'nested' nature of projects, which means that sticking to project plans has symbolic importance, even when it might hinder progress on the innovation itself.

Learning Objectives

The learning objectives for this chapter are to:

1. Differentiate (and see the similarities) between ongoing operational work and innovation projects

2. Understand how projects can stimulate innovation, including digital innovation, through creating distance from ongoing operations

3. Recognise the limitations of project management routines in the context of innovation where emergence is high

4. Understand the nested nature of projects which can account for why project management routines may still be used even when the plans are constantly being redrawn.

Case: SkinTech

SkinTech is a regenerative medicine company that was originally formed in the 1980s as a spin-out from a university. Shortly after the spin-out, the founders commercialised a biological 'Skin' product, short-circuiting a lot of the normal regulatory approvals process because the product was classified as a medical device rather than a drug, for which there was, at the time, far less regulation. The product was marketed as a superior replacement for bandages because it contained biologically active ingredients that could stimulate wound healing as well as protecting the wound. However, as the founders had very little business expertise, they licensed the product to Pharma (a large pharmaceutical firm), which were responsible for sales and marketing. Pharma was interested in the regenerative technology because, at that time, this was seen as having the potential to revolutionise medical treatments, such as eventually being able to grow new organs for transplantation. However, the licensing agreement was dissolved after only a few years as Pharma found that marketing the product involved actively working alongside medics as they learned how to practically use the product, and this was very costly. Coupled with the fading beliefs in the industry about the potential for regenerative treatments in the near-term, Pharma pulled out of the venture.

Moreover, even while Pharma was reassessing its relationship with SkinTech, SkinTech itself was having financial problems because it was costing them more to make the Skin product than Pharma was actually paying them. As a result, SkinTech was making losses and filed for bankruptcy. However, two of SkinTech's founders decided to resurrect it by investing their own money. SkinTech emerged from bankruptcy, but as a much smaller company that was based on a business model focused upon outsourcing rather than extensive engagement in in-house R&D. SkinTech began working with a variety of academic researchers and organisations (e.g., Clinical Research Organisations for conducting clinical trials and manufacturing companies for producing the product) and so moved to more of an open innovation model rather than developing in-house capabilities.

Once the company emerged from bankruptcy, it started to work on other products in order to diversify its portfolio so that it was not reliant solely on the original Skin product. It eventually had three projects that it was working on simultaneously:

1. While the original 'Skin' product, aimed at wound repair, started to become profitable, its market penetration was limited because of the inclusion of animal cells. Thus, SkinTech was working on the next generation of Skin that would be easier for medics to use and would not contain animal products. This project had been ongoing for several years but had not managed to produce an effective substitute for the original Skin.

2. 'Dental Skin' was a second project that was focused on producing a potential dental application of the technology that would support the regeneration of gum tissue. SkinTech was working with dentists to produce this product and was beginning to start thinking about a phase 1 clinical trial.

3. 'Cosmetic Skin' was a third project. This project was focused on producing a potential skin cream using the waste products from the Skin manufacturing process. This waste contains biological material that was thought to be able to regenerate skin and so reduce signs of ageing. Again, it was starting a human trial on this product, which did not require regulatory approval.

Each project was managed by a project leader, with other project members being involved as tasks dictated. Employees were working simultaneously on more than one project. Each project was managed using a variety of project management tools, although one of the project managers noted that 'we have this Gantt chart that's a hundred pages long and I'm constantly tearing it up'. In each project, issues emerged which meant that the project plan had to be amended. Nevertheless, on each project, while the time taken to progress particular activities often overran, the company tried, wherever possible, not to over-run the overall timeline, especially in relation to important milestones, such as the conduct of clinical trials. So, where activities took longer than planned, it then tried to 'squish in' subsequent activities in order to get back on track. We consider progress on each project to illustrate this.

On the Dental Skin project, the dentists with whom SkinTech was working indicated that, given the small area that they were treating (inside the mouth), the existing shape of the Skin product (a circle) was not going to be very effective and they suggested that a square-shaped piece of Skin would be much better. The project team started work to produce a square shape. However, as Skin was a cellular product, it did not easily conform to these requirements – cells grow naturally in a circle as they radiate outwards as they duplicate. The scientists involved tried different ways to force the cells to grow differently, but nothing worked. They then thought about cutting the circular piece of Skin into a square shape. This solution was dismissed both because of product waste – Skin being very expensive to produce – and, more importantly, because there was a risk of contamination from cutting. As they were trying out these different solutions, time passed, and the project plan they were working to indicated that they should be moving to a phase 1 clinical trial. Rather than postpone the trial (and so set the project back and risk the disappointment of investors and the regulatory body that had already authorised the trial), they went ahead using the circular shape of Skin, although they knew this would not be ideal. Perhaps not surprisingly, the trial results were not as positive as they had hoped – the Skin product performed no better compared to existing products used by dentists. This led to the project being de-emphasised in terms of moving to the next stage.

Regarding the Cosmetic Skin product, Skintech also went to a trial in-line with the project plan. In this case, the key problem was finding a formula to embed the Skin waste cells in a lotion that would allow penetration of the outer skin barrier. It tried many different approaches but was not entirely happy that they had solved the problem. Nevertheless, it decided to go ahead with a trial. While in most cases, phase 1 clinical trials are about safety, in this case, as the product was not being ingested and so did not require regulatory approval, safety was not considered an issue and the focus was on efficacy – did this product actually reduce wrinkles? It did not use physicians to conduct trials but worked with a cosmetics company. The trial design was to have participants use the Skin product for a six-week period. They were asked to rub the lotion into their upper right arm. Photos were taken of this area of the body before and after the trial, and photos were also taken of the left arm for comparison. At the end of the trial, the evidence was very disappointing since there was no visual impact that could be discerned. Subsequent tests showed that the lotion had not penetrated the upper layer of the skin, so that the cells in the Cosmetic Skin product could not do any 'work'. Moreover, those conducting the trial were uncertain as to whether their instructions were sufficiently specific – with participants varying in their interpretation of where and for how long they were supposed to rub in the lotion. The disappointing trial meant that ultimately resources were scaled back on this project.

The final project was the new Skin product which SkinTech was trying to re-create but without the animal cells. Work on this was very slow and it struggled to find a substitute. Given the slow progress, it decided to try

Figure 5.1

Clinical research and innovation
© *Cultura*

to increase the market for the existing Skin product by getting regulatory approval in Europe since, at that time, it had regulatory approval to use their product in the USA only. This meant that it had to send a dossier of information to the European regulatory body. However, being less conversant with the regulations in Europe, and as now the regulations had 'caught up' with this new type of biological technology, it miscalculated what was needed and the application was rejected, with multiple problems being pointed out by the regulator. This set SkinTech back further in terms of expanding its market.

We use this case throughout this chapter in order to help introduce, illustrate and discuss key topics (Figure 5.1).

Introduction: Routines and innovation

Much of what goes on inside any organisation can be described as *routine* (Feldman and Pentland, 2003) – individuals are not consciously making decisions every minute about what they should or should not do but rather are following *scripts*, not necessarily scripts that are written down but scripts in some form or other nonetheless. New entrants to an organisation often feel overwhelmed because they are not familiar with the routines given that they have not as yet learnt the scripts, but they tend to pick these up quickly.

This is because there are many social and material arrangements within any organisation that *equip* the context (Gherardi, 2011), so that following the routines appears natural. For example, a cashier may be working in a supermarket, serving customers at the checkout. The cashier does not have to know the price of every item that is sold because there will be barcodes on the items and a scanner at the checkout. In this sense, the checkout area is 'equipped' to help the employee identify what goods are being purchased, how much each item costs and what the total bill should be. Before the advent of this type of digital technology, a shop assistant would need to know or look up the cost of items and calculate manually (or with a calculator or cash register) the total bill. So even in the past, the workplace was equipped to help the employee as there could be a sheet of paper with prices and probably some means of calculating the total bill.

With the advent of digital technology, the 'equipping' of the context to enable a worker to carry out their daily tasks is much more comprehensively developed. As a result, today, digital technology is a major part of this equipped context, 'scaffolding' (Orlikowski, 2006) everyday practices in the sense of enabling and constraining what people do (e.g., the barcode and scanner do not allow a cashier to modify the price of an item unless they ask a supervisor to override the system-generated price). We can also see how a person involved in purchasing decisions follows a digitally enabled process when ordering supplies for the organisation, or a university lecturer uses various tools, such as a computer and projector, to present materials to the class. Similarly, a doctor creates a patient record using a pro forma, often directly on a computer or some other kind of mobile device, checking off certain symptoms, making a diagnosis and organising a treatment plan, ordering tests, and sending messages to other healthcare professionals involved in the patient's care. Of course, each of these actors has some starting knowledge specific to the particular job, but the actual way they carry out the job follows a 'natural' course, based on established routines in the particular context of work that is equipped to support these routines.

Any given individual undertakes only a limited number of tasks – and so follows only a limited number of routines. Routines interface with other routines, however, in order to produce whatever it is the organisation does, whether this relates to products and/or services or both. We described this as a site of practice in Chapter 4. In this way, routines (repetitive actions, generally reflecting underlying processes undertaken by employees) relate to the organisation's knowledge (not individual knowledge). This allows organisations to use routines, such as to pursue an organisational aim, in ways that are not necessarily associated with the people who actually perform them. This is why Nelson and Winter (1982) argue that it is companies that know how to make things, not people.

Routines also suggest a degree of rigidity and inertia – repeating things that have been done in the past. This is the case even though it is recognised that enacted routines do not always or even mostly follow the established script. This is why Feldman and Pentland (2003, p. 94;101) made the important distinction between the *ostensive* aspect of routines – the 'abstract, generalized idea of the routine' – that 'enables people to guide, account for, and refer to specific performances', and the *performative* aspect of a routine – the 'specific actions, by specific people, in specific places and times' – that 'creates, maintains, and modifies the ostensive aspect of a routine'. In Chapter 4, we described this as the difference between practice and praxis, and along the same lines, Pentland et al. (2011, p. 1369) demonstrated that even 'routine routines' (in their case, processing purchase orders) exhibit considerable variation.

While routines support everyday organising, it may seem that the notion of routines is not applicable in regard to innovation. After all, the essence of innovation is the production of something new, so following routines may seem like the antithesis of this objective. Nevertheless, routines are still used to support innovation – in particular, routines around the management of projects. Indeed, there is a major industry targeted at teaching individuals the routines of project management – how to produce project plans and Gantt charts that set out when tasks will be started and completed and by whom, identifying risk, monitoring progress, conducting milestone reviews and producing a post-project review of lessons learned. Each of these tasks is today supported by a plethora of tools (often digital) that guide a person in managing a project, although the project itself may well be focused on producing something innovative.

There is a potential paradox here – when one is trying to produce something new and different, it may be difficult to know (and so plan in advance) what tasks will be done, at what point in time, and by whom and over what time period. Bessant et al. (2011) recognised this paradox and argued that the 'trick' for those firms that are in the business of innovation is to develop routines that allow regular innovative activity, so that firms need to simultaneously (i.e., along with following project management routines) acquire the ability to review their routines and change or modify routines as required, so that they can cope with the unexpected. This again refers to the ambidexterity idea of exploiting new routines while exploring new ways to do things; ambidexterity, as we have seen, is challenging yet rewarding in terms of a firm's performance.

We first look at how projects are supposed to support innovation and then consider some of the associated problems. In examining these problems, we also consider how the type of innovation (whether complicated or complex – the distinction is given below) may influence the extent to which traditional project management routines are applicable. We end by considering the nested nature of projects, which illustrates how project management has a symbolic, not simply functional, role. Before we do this, we differentiate between projects, teams and teaming.

Projects, teams and teaming

An overview of projects

Projects are typically used to organise activities associated with developing or introducing innovation, as in the SkinTech case. Projects are therefore initiated to accomplish pre-specified goals and objectives, such as to develop a new type of electric car or some new organisational service. Project (teams) are given at least some autonomy – somewhat separated from ongoing organisational activities – so that the people involved are unencumbered by established routines and practices. We discuss this further in Chapter 6 where we draw on the concept of *liminality*. Given the upfront purpose of projects they are different from teams per se, however, in the sense that projects involve more 'coming and going' of individuals; projects rarely involve a stable group. This is especially so with innovation projects since these can often involve a very long timeframe – as in the SkinTech case and as with ongoing attempts to develop autonomous vehicles. Indeed, a specific innovation may involve not just a single project but multiple projects; Grabher (2002) describes this as a *project ecology*. In a similar vein, Desouza and Evaristo (2004) distinguish three types of project context in addition to stand-alone projects. First, there are projects that are part of a *co-located* programme; here, multiple projects run concurrently at one location. An example would be multiple projects, each working on a different aspect of the development of an autonomous car – with some projects focusing on the interior design while others focus on the sensors and algorithms to allow the car to self-navigate.

Second, there is the *distributed* project, which is a single endeavour conducted from multiple locations, often involving different organisations. Our car example could be organised like this as companies move towards more open modes of innovation (e.g., Chesbrough, 2003) and involve different partners who each focus on their specific part. We discuss this more fully in Chapter 6. Finally, there are *multiple* projects taking place at multiple locations and often distributed in time – a context that can be described as a complex project 'ecology'. An example would be the development of a new drug, which may start life in a research lab of a university, where a new molecule involved in some disease is identified. To translate this breakthrough science into an actual new drug is likely to involve multiple other projects in different organisations over an extensive period of time. Indeed, the average time taken to develop a new drug is 14 years (Van Norman, 2016). This latter type (i.e., multiple projects), Desouza and Evaristo (2004) argue, is most difficult because it requires the management of multiple (time/space) interdependencies across projects. Newell et al. (2008) demonstrated the difficulties of managing interdependencies when projects are part of a complex project 'ecology' – that is, where there is a broad range of projects (not necessarily benefitting from physical or timing proximity so that multiple partners are involved over an extended period) involved in some innovation initiatives. We shall deal with this context later when we discuss differences between complicated and complex projects. SkinTech provides an example of such a complex *project ecology* because the original Skin product was 'invented' in a university research lab several years prior to their current project to replace the animal cells in the product. Many digital innovations similarly start in one place – perhaps university research labs – and then get picked up by different types of organisation over time and space.

Projects and teams

Teams are traditionally defined as groups of individuals who share a common identity and so contribute to the collective (team) good rather than (at least theoretically) their own personal advantage. However, individual team members might of course get something out of the team doing well. This is an *ideal type*, and even in what we might describe as an archetypal team such as a football team, we often get acts of individual selfishness, as when a player tries to score a goal rather than pass the ball to a teammate who is in a better position to do so. Moreover, we can also see acts of 'collectiveness' in groups of strangers – as when a bystander helps someone in trouble, even though this may be inconvenient and even dangerous.

In projects, there may be an overall project goal; however, individuals may well be involved only intermittently, performing a specific task, without necessarily having much knowledge or interest in the overall project objectives. Nevertheless, even in projects where individuals are assigned specific tasks according to the project plan, there is a need for people from different backgrounds to engage in some kind of collaboration – even when they are involved only temporarily. After all, that is the *raison d'être* for a project; a group of homogenous people – who all think alike (were this possible) – is unlikely to produce new ideas and practices because there is nothing to stimulate alternative thinking. So, projects bring together people with different knowledge, skills and ideas because innovation is more likely to occur at the boundaries between disciplines, departments or cultures. However, project personnel have to collaborate and overcome knowledge boundaries if they are to succeed (see Chapter 6 on the role of knowledge brokers). Edmondson (2012, p. 74) uses the term *teaming* to describe what is needed in these situations. She describes how, in order to build what many consider to be a spectacular structure – the Water Cube for the Beijing Olympics – 'dozens of people from 20 disciplines and four countries collaborated in fluid groupings' (Figure 5.2).

Project and teams: Principles for success

Edmondson (2012, p. 75) describes the success of this project as being the outcome of effective teaming and 'a way to gather experts from far-flung divisions and disciplines into temporary groups to tackle unexpected

Figure 5.2

Grouping is key for facing team-based challenges
©Getty Images/iStockphoto

problems and identify emerging opportunities'. She argues that teaming depends on embracing two sets of principles:

1. *Project management*, including predefining the project scope, structuring activities and arranging tasks depending on their levels of interdependence, and

2. *Team leadership*, including ensuring there is clarity of purpose, building psychological safety and embracing conflict and failure.

These are also seen as central ingredients of effective project management. We consider this next as we critically review research related to projects and innovation.

Project management

A project 'team' is typically composed of representatives from different departments. The project will be a learning experience for this collection of people who will need to creatively explore how to produce a new type of product, service or organisational process (i.e., to innovate). Therefore, the

Key Concepts: Routines and Projects

Routines:

- Scripts – not necessarily written down but known ways to carry out a set of tasks
- Equipped context – the physical arrangement of tools that make it easy for someone to follow the routine
- Ostensive and performative aspects – the script as it can be described (ostensive) versus the routine as it is actually enacted (performative)

Projects:

- Co-located, distributed and multiple projects – projects can be more or less complex
- Project ecology – the overall distribution of a project over time and space
- Teaming – a way to bring people together to collaborate quickly in the context of a project

very essence of a project is that it is separated from the rest of the organisation. As Engwall (2003) noted, projects can be a 'lonely phenomenon'. Nevertheless, the project team must ensure that those who will eventually be adopting 'their' innovation will accept and learn and understand how to use it as well as accept any organisational changes that the project team is endorsing. This need for mutual connectivity between the learning within the project team and the learning across and beyond the organisation (e.g., including potential adopters) can be challenging. This is so because projects, by their nature and design, are, to varying degrees, independent from the rest of the organisation and other stakeholders. Indeed, it is this relative autonomy and the degree of decision discretion that this confers on project members that promotes more creative and innovative solution development since it allows project teams to respond more flexibly and speedily to external demands, unconstrained by the normal hierarchical and functional boundaries (Gann and Salter, 2000) that restrict communication and interaction in bureaucratic organisations. At the same time, however, Swan et al. (2010) demonstrated how this isolation between the project and the wider organisation can have a detrimental impact on knowledge transfer from the project to the wider organisation, describing this as a *learning boundary* between the project and its various stakeholders – we saw this in the TechCo case in Chapter 2. We discuss this also in Chapter 6 through introducing the concept of *liminality*, as previously noted.

Aside from this learning boundary, there are more fundamental criticisms surrounding projects. Projects are presented as an alternative to bureaucratic forms of organising. Indeed, as we discussed in Chapter 3, while bureaucratic forms of organising are said to be effective for stable situations, projects are heralded as a necessary 'new' form of organising given dynamic environments which require more flexible and versatile forms of work organisation (Meredith and Mantel, 2011) – and, as a result, greater innovation. In this way, a project-based form of organising is contrasted to a bureaucratic form of organising, with projects enabling flexibility in a way that bureaucracies do not. The Project Management Institute (the largest professional body for project managers), in its guide to *The Project Management Body of Knowledge*, defines a project as 'a temporary endeavour undertaken to create a unique product, service, or result' (https://www.pmi.org). This definition suggests that projects are the best mode of organising to enable innovation.

Project characteristics

In terms of the characteristics of a project, they are provided by the PMBOK (project management body of knowledge) guide, which summarises best practices, technologies and guidelines, which are accepted as standards within the project management discipline. The guide suggests that two key features of a project are its *temporary nature* and its focus on creating something *unique*. These aspects of the definition have remained consistent since the PMBOK was first published in 1996. However, in later editions, while the same definition is provided, the guide actually goes on to describe three characteristics of projects, with *progressive elaboration* added to the temporary and unique characteristics stated in the original. The guide suggests that progressive elaboration means that projects will be elaborated incrementally over time but also advises that this progressive elaboration should not be confused with 'scope creep'. So, the scope of the project – the objectives to be met – should be predefined to ensure that the project can be calculable, although project specifications can be progressively elaborated. As Thomas (2006, p. 92) reflects, 'this addition to the definition appears to be almost an afterthought that does not appear to influence the rest of the material in the PMBOK guide'. Instead, the bulk of the guide is focused on presenting tools and techniques that will enable the accurate planning and control of projects. Such a focus fits with mainstream management discourse in general, which is concerned with defining how to make activities undertaken in work more predictable (i.e., routine) and controllable through the use of objective, quantitative techniques that are based on an instrumental rationality (Wood, 2002) (Figure 5.3).

In terms of these techniques, following from this definition of a project, the PMBOK guide (2004, p. 8) goes on to define project management as 'the application of knowledge, skills, tools and techniques to project activities to meet project requirements'. These tools and techniques are applied to the project management processes of 'initiating, planning, executing,

Figure 5.3

A word cloud showing where emphasis in project management typically lies

monitoring and controlling, and closing'. Therefore, managing a project involves a series of phases, such as identifying *requirements*, establishing clear and achievable *objectives* and balancing the *triple constraint* of scope (what will be achieved), time (how long it will take) and cost (how much it will cost). Again, in the description of these processes, the PMBOK guidebook (2004) adds a fourth dimension – *adapting* 'the specifications, plans, and approach to the different concerns and expectations of the various stakeholders', but again this adaptation needs to happen while avoiding the evils of scope creep. This is to be avoided because scope creep will usually involve enlarging what the project can deliver in terms of functionality and such creep will inevitably increase time and cost.

The changes in the PMBOK guide reflect the evolving discourse on project management. Whereas early project management prescriptions focused on the technical aspects of management, later developments included the human aspects. Nevertheless, these more 'intangible' human elements are still treated as if they can be controlled in order to ensure predictability (Townley, 2002). For example, critical success factors have been identified that supposedly help to ensure that the 'fickle humans' involved in projects remain committed and act consistently in fulfilling the tasks that they have been assigned. So, aside from the technical control tools such as Gantt charts and PERT (programme evaluation and review technique) that are to be used, it is also important to have, for instance, top management support so that those involved will recognise the value of what they are doing and so focus their efforts on achieving the objectives and user involvement so that 'users' will willingly adopt the product of the project, whatever this may be (Figure 5.4).

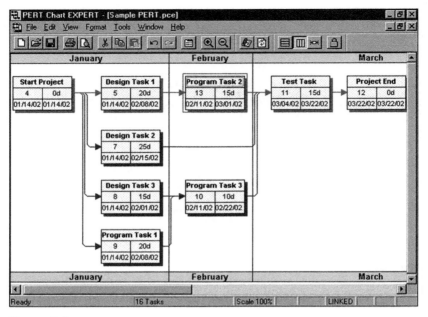

Figure 5.4

An example of a PERT chart in the software WBS Schedule Pro (screenshot reproduced with permission from WBS)

Effective project management

While considerable energy has gone into defining the 'how to' prescriptions for effective project management, Cicmil and Hodgson (2006, p. 2) nevertheless conclude that 'There is little evidence that the resulting torrent of competing streams of thought, methods of enquiry and best-practice claims and propositions has creatively contributed either to constructive debate in the field or to resolving the difficulties encountered in practice'. Indeed, in terms of practice, evidence suggests that, despite the growth of sophisticated project management methodologies, only very basic ones actually get used and, even then, often not as intended (Besner and Hobbs, 2004). Thomas (2006) goes as far as to say that the mainstream project management literature is best described as having its 'head in the sand', promoting an idealised concept of projects and project management, at the expense of 'the lived experience of projects'. Nevertheless, project management tools and processes discipline work and workers, even while they may fail very often to ensure that projects achieve their stated objectives on time and in budget. Indeed, it is the attempt to discipline work, which is inherently indeterminate, that creates many of the problems in projects. We discuss this further when we consider the distinction between complicated and complex innovation projects in the section that follows.

If we unpack what is described in the mainstream project management literature, it is arguably very similar to the discussion of bureaucracy in the sense that it tends to rely on output control (prescribing what should be done by when) and discipline (finding people 'at fault' when things are not done to plan). This is the case even though project management is often depicted as different from operations management. The PMBOK guide (2004, p. 7) states that 'the objectives of projects and operations are fundamentally different. The purpose of a project is to attain its objective and then terminate. Conversely, the purpose of an ongoing operation is to sustain the business.'

However, the same emphasis on control is clearly evident in the mainstream discussion about project management as it is in relation to operational control. The tools, techniques and methodologies that have been developed purport to ensure that projects will be delivered on time, in budget and to scope, providing, of course, that the tools, techniques and methodologies are 'properly' utilised. This led Townley (2002, p. 564) to describe mainstream project management prescriptions as 'a series of carefully delineated steps, executed in sequential order: optimal decisions are made by careful planning and rational logic'. The assumption is that progress on a project will occur in a predefined, linear way so that work can be divided into small, measurable units that can be closely monitored. This is reminiscent of a Scientific Management approach to work, which assumes that it is possible to rationally plan and execute social activities (see Chapter 1). As Thomas (2006, p. 103) concludes, project management 'does not resolve the issues surrounding the dysfunctional aspects of bureaucracies; it merely changes the scale of the operation', and (p. 100) the aim of project management tools and techniques is to 'enlist more of the team members' labour by enrolling the self-identity of the worker so that they subjectively define their worth in terms of their ability to contribute to the success on the project'. Lindgren et al. (2006) go as far as to call projects 'mental prisons', being even more controlling than traditional bureaucratic structures.

We can summarise the problems associated with the routines of project management in relation to the *temporal* and *spatial* features of projects. Relating to their temporal features, projects are often defined in terms of increasing flexibility. However, this flexibility relates to setting up a project to do something different; it does not refer to the flexibility to actually accomplish this 'something different'. Instead, as we have seen, mainstream project management is essentially about predefining and controlling the activities needed to accomplish this 'something different' so that outcomes are predictable (with project team members simply following the plan). Thus, rather than increasing flexibility, project management tools build in rigidity in a way that is antithetical to the needs of interactive innovation (see Chapter 2). Moreover, the division of activities into distinct temporary projects may make it difficult to think strategically about the long term – with the risk that projects will be just 'isolated sequences of action lacking any meaningful links to both the context and the future' (Lindgren et al., 2006 p. 113).

In terms of the spatial features, while project management methodologies do increase horizontal communication and coordination within the confines of the project, they also separate the project from the rest of the organisation and so create 'blinders' (Thomas, 2006) rather than the openness that is needed for interactive innovation. Thomas states that project management concepts and standards based in linear, rational, controlled systems provide a certainty that removes ambiguity and makes us more comfortable. This desire 'to be comfortable', to Thomas, leads us away from uncertain and uncomfortable, 'double-loop' learning processes that enable us to deal with complex adaptive systems. Rather than helping us to manage these complex processes, then, the mainstream portrayal of project management attempts to reduce the complexity through simply ignoring it. But, of course, putting one's head in the sand does not make the danger go away. Further, from the project workers' perspective, projects can also be lonely, requiring working long hours, often in remote locations that can disrupt family and workplace relationships.

Interactive innovation recognises interdependencies between activities but is more sceptical about how easy it will be to pre-plan this sequence ahead of time because, by definition, the results of at least some activities are not able to be (accurately) predicted in advance. This is because, when one is dealing with many innovations, there are likely to be a number of resistances that are encountered along the way that must be accommodated (Pickering, 1995) through negotiation (Wagner and Newell, 2006). Since the resistances are not known in advance, the negotiations and resulting accommodations will be emergent. We discuss this further as we consider differences between complicated and complex innovation projects.

Types of projects and project management

The extent to which an innovation project can 'fit' into project management routines discussed above may depend on the type of innovation that is involved. Some innovation projects may be *complicated* but, nevertheless, it may be possible to know in advance what needs to be done and, even though problems may emerge, solutions can be readily identified if experts with the necessary knowledge are involved. On the other hand, Dougherty and Dunne (2012) argue that some types of innovation may be described as *complex*, and these types of projects pose particular management challenges, relying on the kind of complex project ecology discussed earlier. The difference between a complicated versus a complex project is illustrated by the difference between manufacturing a new variant of a human-driven car versus a self-driving (autonomous) car. The new human-driven car may have novel features planned that may be more complicated to design than originally anticipated but solutions are ultimately going to be found. The new self-driving car brings in a new order of magnitude of complexity because we have not yet got the digital technology to support such a car, at least a car that will work in 'normal' conditions on busy roads used by other, non-self-driving units, such as pedestrians and human drivers who may be more or less distracted. The SkinTech project can be described as complex, for example, because there are unknown aspects to the science, including growing cells in squares rather than circles for the dental application.

Complicated and complex innovation projects

Both complicated and complex innovations involve multiple actors who must draw together their distributed knowledge in order to create the new product, service or organisational arrangement. For this reason, both types of innovation involve considerable interaction between actors. However, in complicated innovation (such as designing a new petrol car), knowledge of the kinds of problem likely to be encountered and cause–effect relations mostly already exist. Ultimately, provided that appropriate solutions can be proposed by the experts, alternative courses of action can be predicted and evaluated (Dougherty and Dunne, 2012). In complex innovation processes, problems and cause–effect relationships are largely unknown or are still being worked out (Styhre, 2006; Dougherty and Dunne, 2012) – as in our examples of the self-driving car or the square piece of Skin. Complex innovations are characterised by very high levels of uncertainty, even *unknowability*. Unknowability means that it is not possible to know what will be the outcome of project tasks that must be undertaken. For instance, in relation to our self-driving car, we don't have the full range of technologies that will ultimately 'work' nor know how much we will need to invest in both the car and the infrastructure before we develop the breakthrough technology. That is why the companies working on self-driving vehicles are those with very 'deep pockets'. As another example, for a new drug project, even after extensive pre-clinical work, when a particular therapeutic substance in laboratory and animal experiments seems to perform well, when it comes to scientists beginning to work with humans, the results can be completely different. Complex innovations also typically span many years and require significant financial investment without any guarantee that a commercial new product or service will be produced – this is why so many small biotechnology firms trying to develop new drugs and many digital companies trying to develop new technologies have gone bust. In this sense, these examples are 'radical' innovations (see Chapter 2), which entail substantial risk, significant time and cost, and the continued discovery and development of new science/ knowledge.

Dougherty and Dunne (2012, p. 1467) summarise the characteristics of complex innovation as 'nonlinearity, unpredictable interdependencies and the emergence of knowledge over long periods of time as innovators search the unknown unknowns'. In other words, complexity exists in situations where we don't yet know what works and where unanticipated results can emerge at any point and impact other activities related to the project. For instance, new data about the toxic effects of an experimental drug during clinical trials may mean that new research is needed before further trials can be conducted, making it impossible to meet the plans for the next scheduled set of trials. And in relation to self-driving cars, we have not yet established whether we can agree on ethical criteria (e.g., hit a child or an elderly person) to program a car when an accident cannot be avoided. This example illustrates not only the problems of trying to plan out the schedule of trials in advance, before results from one set of trials can allow the manager to know whether the next trial can go ahead; it also illustrates the interactive nature of innovation, with research having to be done after development has started rather than all development following all research (Figure 5.5).

Complicated and complex innovation development

The illustrations of medical product development provide good examples of a complex innovation context, as does the example of a self-driving car. In this type of complex context, different scientific specialists must work closely together in order to achieve the scientific breakthroughs needed to develop a radically new technology. Moreover, this specialist scientific knowledge

Figure 5.5

Development of innovative drugs
©Getty Images/Westend61

needs to be integrated with commercial, management and product development expertise in order to secure the resources needed to progress development. In the early phases of each type of development, the underpinning science will often be only partially understood. Since much is still unknown about the biological or digital mechanisms underpinning human disease or self-driving infrastructures, animal testing or simulation can often not predict how a novel solution will work. This means that the development process is neither linear nor smooth (Swan et al., 2007).

At the same time, this type of breakthrough innovation work is often done in small firms that are trying to exploit some new science. They will often be working with an array of other organisations to help them progress the development (e.g., small biotechnology firms typically do not conduct their own clinical trials but outsource the conduct of these trials to specialist trials firms, as did SkinTech). At the same time, these small firms often realise that they will not be able to take the innovation all the way through to commercialisation since they do not have the resources to do this. Their objective instead is to show enough development progress to entice a larger firm to licence their technology or simply acquire their company. The larger company will then have the resources needed to continue with the uncertain (but now slightly less uncertain) development. For the larger firm to be attracted to the smaller firm that is developing the breakthrough technology, it needs to be convinced that the development process is progressing well. Small firms will therefore publish the fact that they are developing the new technology together with a clear statement of when they believe the technology will be market-ready. These pipeline statements are often published on their website as evidence that they are in control of the development. It is in this type of complex project ecology that we can begin to understand how project management methodologies (including the publication of pipelines and associated project management plans) have symbolic rather than simply functional significance – an issue to which we now turn.

Sources of uncertainty in projects and project learning

Given the high level of uncertainty – sometimes *unknowability* – in complex project contexts, Rice et al. (2008, p. 54) advocate the use of *learning plans* in place of project plans. They argue that learning plans are more suitable in contexts of so much uncertainty; 'for any breakthrough innovation project, specific objectives are often unclear or highly malleable, and the paths to them are murky. Rather than feign a certainty that doesn't exist, project managers need a systematic, disciplined framework for turning uncertainty into useful learning that keeps the project tacking on a successful course.' They describe four main sources of *uncertainty* in breakthrough projects – *technical, market, organisational* and *resource* uncertainties. We provide examples of these as we discuss some of the problems associated with implementing a learning plan.

The idea of the learning plan is that managers should identify particular sources of uncertainty in relation to each type of uncertainty, propose alternative assumptions related to this uncertainty and then find ways to quickly test the assumptions so that the uncertainty can be resolved. Rice and colleagues provide an example of a learning plan in an innovation project that involved the development of a hybrid electric delivery truck. There was uncertainty regarding what were the needs of delivery truck drivers. As a result, the project team actually experienced a 'day in the life of' a truck driver. This led them to realise the importance of reliability and they subsequently worked with a reliability consultant to 'harden the design for real-world use' (Rice et al., 2008).

While the learning plan may be a potentially useful idea, it is also the case that, in more complex innovation contexts, sources of uncertainty may be more difficult to root out quickly. For example, until a clinical trial is actually conducted, the safety and efficacy of a particular compound in the human body cannot be known with certainty. As we know, there are cases where the side effects of drugs are finally established only after a drug has been used for several years. In the thalidomide disaster, it took several years before it was finally determined that this drug was causing serious deformities in children. Take a look at the 2012 article in the *Guardian* newspaper entitled 'Thalidomide scandal: 60-year timeline' for more detail on this. These side effects may necessitate the removal of a drug from the market and the loss of any potential revenue. This may be so, even after several million dollars were spent on its development; the average drug costs about USD $1 billion and is 14 years in development (Van Norman, 2016). The e-waste problem created by the digital revolution is another example of how innovations create problems, down the line, that have not been anticipated and costed-in by the innovators, leaving societies struggling to pick up the costs both financially (e.g., actually disposing of the waste) and socially (health issues created by toxic fumes) (Figure 5.6).

Much of the mainstream popular literature suggests that, even in such complex innovation contexts, traditional project management tools can and should be used as part of the routine of organising of such activities (Sheremata, 2000; Styhre, 2006; Andriopoulos and Lewis, 2009). However, while such project management tools and techniques are used and can be useful, it is also the case that research in these complex project contexts suggests that those involved regularly encounter problems that delay activities and so mess up the neat project plans that are repeatedly prepared. However, in the face of this, while those involved do 'tear up the Gantt chart' (as in SkinTech), they then proceed to produce a new plan (and associated chart), attempting to work around the obstacle that has 'messed up' the initial plan. This means that, in producing a new Gantt chart, a main aim is to squeeze in tasks that

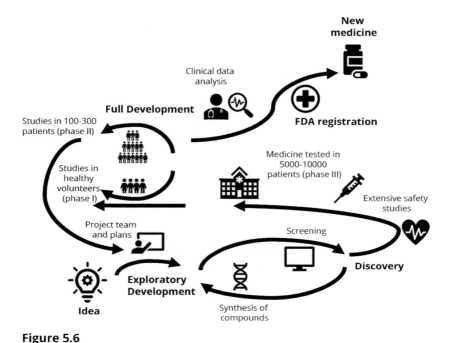

Figure 5.6

The long road to a new medicine

have been held back so that the major milestones agreed for the project at the outset are not derailed. SkinTech did this by going ahead with a clinical trial with dentists even though it had not yet managed to produce the square shape that dentists had asked for.

Four types of uncertainty in innovation projects

We can identify four types of uncertainty in innovation projects. These are outlined using the framework developed by Rice et al. (2008) in order to illustrate the different types and also the sources of uncertainty in the context of drug development innovation projects:

1) *Technical uncertainties:*
   Given that the science is still evolving and the huge costs of drug development, firms try to protect the science they are using. In drug development, this means taking out patents on particular areas of the science (or the methods being used to develop the particular therapeutic drug) – this is referred to as *Intellectual Property* (IP). Firms may seek patent protection on their own IP or they may seek to license or buy IP from others. One firm that we researched, in an attempt to protect IP in a particular disease area, purchased some IP from another firm – IP that it hoped

would help it develop a new therapeutic. In the event, despite undertaking due diligence on the IP, it turned out that the IP was much less useful than the firm had anticipated because the firm would have needed to do a huge amount of additional work on the science before a new drug could be proposed for initial regulatory approval for first-stage clinical trials (in the words of the CEO, the IP turned out to be a 'fake painting' rather than 'the Rembrandt' that the firm had anticipated) (Newell et al., 2008). This resulted in a lot of waste – of time and money – and in the end it abandoned the project. In relation to digital innovation, there are also IP issues but there is also an increasing trend to create open-source software that overcomes some of these issues.

2) *Market uncertainties:*
   Given the high cost of drug development, firms need to be able to recoup their outlay by selling the product, once on the market, at high cost. Some have argued that the high costs are inflated by inefficiencies in the industry itself. Indeed, there are often considerable amounts of wasted resources, as we have highlighted in our examples. Additionally, there are marketing costs. Therefore, it is the case that a firm needs to make a sustained profit over time. This issue of 'pay back' is especially the case in relation to drugs that are developed for relatively small populations (e.g., for diseases that are rare). To overcome this problem, in healthcare contexts, where insurance companies foot the bill for therapies, drug companies spend a lot of time convincing insurers that their new drug is substantially better than existing drugs. However, in countries where there is a national health service, this is more difficult and there are several examples where, in the UK for example, the National Institute of Clinical Excellence (NICE) – the body that approves new drugs – has turned down the use of a new drug because of the expense as compared to the perceived benefits. This then reduces the market opportunity for the drug company. The problem is that it is very difficult to foresee what NICE (or its equivalent in other countries) or even insurance companies will approve. In the SkinTech case, we saw how the bid to use its Skin product in Europe was turned down by the regulator. In relation to digital innovations, there is also a lot of market uncertainty, especially with more radical innovations where even seeming successes can quickly turn sour, as was the fate of Napster and MySpace and many others.

3) *Resource uncertainties:*
   In one project we have researched, the team were focused on developing a new type of obesity drug that would not only inhibit fat ingestion (this type of drug, called a lipase inhibitor, is already on the market) but also combine the non-digested fats into a solid. Specifically, they were using a polymer to prevent the 'leaky waste' problem that is an unpleasant side effect of currently available fat digestion-inhibiting drugs. The project team were therefore trying to manufacture a drug solution that combined

a lipase inhibitor with a polymer. They had to outsource this since they did not have the manufacturing capability themselves and there was a significant lead-time on manufacturing. This meant that they had to manufacture at least nine months ahead of any projected clinical trial. However, ahead of a scheduled clinical trial, they were still conducting pre-clinical work in order to determine the ratio of lipase inhibitor and polymer. They therefore faced a dilemma – produce the solution ahead of time, before there was certainty about the ratio, or wait until this was certain but then have to delay the clinical trial because manufacturing would take time. They decided to go ahead and manufacture the solution for the trial. However, after some additional pre-clinical work it was decided that the ratio needed to change so that the already manufactured solution had to be scrapped at a cost of several million dollars. Moreover, new material then had to be manufactured, further delaying the trial – the very problem the team had been trying to avoid. In relation to digital innovations, it is often not simply the resource uncertainties relating to developing the technology that must be borne in mind but also creating the infrastructure for the product or service. For example, the resource uncertainties for the self-driving car we have discussed at various points in this book are related not only to developing the car itself but to the resources that will be needed to create the infrastructures and the regula-tions that will make this technology successful.

4) *Organisational uncertainties:*

The fourth source of uncertainty relates to organisational issues. Above, we pointed to the fact that in many breakthrough innovation contexts it is small firms that are doing the initial science and development work. Small firms are always more vulnerable to economic and other external conditions. This means that these firms can be acquired, merged or sim-ply 'go bust'. This type of organisational uncertainty means that decisions are often taken that are short-term focused. An argument often heard might be along the lines of, 'Let's get the next clinical trial done, hope the results are good, and perhaps we'll get some more capital to move to the next stage of development, for example, through some venture capital or an acquisition'. In this type of uncertain organisational context, the 'long game' may not be the most important focus, leading to suboptimal deci-sions just so as to be able to continue to exist. The firm developing an obesity drug mentioned above, for instance, finally conducted a clinical human trial before it had been able to manufacture its drug in a pill form. It therefore used the liquid form of the drug. However, because this tasted horrible, it mixed the drug liquid with apple juice in order to mask the taste. The problem was that this also meant that when the results were not as good as the firm had expected (the problem of leaky waste was still very apparent), it did not know whether this was because the drug did not work or because of the copious amounts of apple juice people had

had to consume. The SkinTech case also illustrates how squeezing in a suboptimal clinical trial can create problems when results don't turn out as expected. In relation to digital innovation, we have seen the example of Kodak, where it was not the problem of developing the digital camera that was the biggest challenge but of convincing others in the organisation that they needed to adapt their existing business model.

The nested nature of projects: Projects in their institutional context

These examples raise the question of why those involved don't just 'tear up the Gantt charts' and leave it at that. Instead, as we have seen, they often repeatedly rewrite them in an attempt to get projects back on track. In other words, why do project managers simulate a certainty that doesn't exist in these types of project contexts (Rice et al., 2008)? The question is especially pertinent when this 'certainty' often leads to squeezing things in that shouldn't be squeezed in, which actually then sets the project back further – such as when a clinical trial fails because it was using what was known to be suboptimal technology (as with mixing the drug with apple juice or using the round shape of the Skin product in the dentist trial).

One way of thinking about this is to realise that projects do not exist in a vacuum. Rather, we can view any given project as nested within a broader organisational and institutional context. Within any organisation, a particular project must fight for resources, often against other innovation projects as well as against ongoing operations. Projects that appear to be missing important deadlines are more likely to see their budgets threatened or withdrawn. Even more importantly in the context of complex innovations, which are often undertaken in relatively small firms (as we have seen in relation to drug development or digital innovation), it is important that outsiders, especially stakeholders who provide investment for continued development, see progress being made. Yet progress is very hard to demonstrate when there are so many unknowns. The ability to demonstrate progress may be different in complicated versus complex innovation contexts.

For example, in relation to digital innovations – a complicated innovation setting – there are well-known problems of keeping projects in scope, on time and in budget. Such projects are often very large undertakings (e.g., developing an integrated enterprise system for a large globally distributed firm, as in the TechCo case in Chapter 2) involving multiple subprojects that are often geographically dispersed (e.g., with testers in India or Vietnam and systems analysts in the USA or Europe). With a traditional project management methodology, all aspects of such a project would be planned in advance and then (theoretically) executed to a pre-specified timetable. However, often things do not go to plan for a host of reasons since, whether the context is complicated or complex, there are still many sources of uncertainty.

However, there are ways to overcome this uncertainty in complicated contexts. For instance, in relation to digital innovation, today more and more such projects are moving away from the planned – or waterfall – methodology to what have been described as 'agile' methodologies (e.g., Desouza, 2006). In agile methodologies, small parts of the software are developed, tested and rolled out. Problems with each small piece therefore emerge very quickly and it is not necessary to wait until the end to realise that the technology is not supporting users in ways that are effective. Specific examples include use of Scrum and Kanban methodologies, and in many ways, such rapid methodologies are aligned to the learning plan as advocated by Rice et al. (2008) (Figure 5.7).

In complex contexts, innovations such as rapid development, as with learning plans, are difficult to execute. So, for example, it is hard to produce a prototype of a new drug and certainly not one that can be 'tried out'. As we have already discussed, some drugs are not seen to fail until they have been on the market for several years, since adverse side effects may take time to appear and may affect only a small proportion of people taking the drug. In the absence of more visible signs of progress, progress against the milestones set out in the communicated pipeline of a firm's drugs in development takes on special symbolic significance. Institutional theory has long alerted us to the idea that organisational practices are often symbolic rather than functional, or as Meyer and Rowan (1977, p. 343) indicate, they are 'rationalized

Figure 5.7

Challenges of software innovation
© Getty Images/Cultura RF

myths' – ceremonial acts that are geared towards obtaining legitimacy rather than actually fulfilling any substantive purpose.

This example helps to explain the continued use of project management methodologies in these contexts of complex innovation, even while this creates the paradox that attempting to stick to deadlines can actually set back, rather than encourage, movement towards successful innovation. Therefore, in the example of the obesity drug above, manufacturing the solution ahead of time in order to stick to the plan actually created additional problems in the end because it wasted resources and led to even further project delay. The same is the case with the clinical trial involving apple juice and the use of the circle of Skin in the SkinTech case. In other words, although project management routines that encourage planning, monitoring and controlling project work in ways that assume a linear progression can actually be counterproductive, such routines have a symbolic role in demonstrating progress in the absence of more concrete information about an innovation's development. Perhaps this explains why so many big companies that are developing fully self-driving cars have clear dates when they expect the technology to be ready, even though the infrastructure and regulatory issues that will need to accompany this technology seem to be far behind.

 Key Concepts: Problems with Project Work

Learning Boundary:
The boundary that can arise when the outcomes of a project are rolled out across an organisation because members of the rest of the organisation have not been on the learning journey that project members have, so that they don't necessarily understand what is being proposed nor why

Flexibility:
Projects often need to be flexible because unexpected things crop up that need to be dealt with; project planning methodologies, on the other hand, assume that everything can be planned in advance

Complicated versus Complex Projects:
Some projects are complicated because there are many diverse activities that need to be coordinated, but with good planning this is manageable; complex projects involve unknowns that it is more difficult to plan for, making them more difficult to manage and the outcomes more uncertain

(continued)

Sources of Project Uncertainty:
- Technical – the technologies that are being developed or used in a project
- Market – the market potential of the product or service being developed in a project
- Resource – how much and what types of resources will be needed to progress a project
- Organisational – where support and opposition for a project will come from

Symbolic Progress:
Project management tools and methods are used even in complex projects where such planning is unlikely to be realistic. We can understand this if we consider how these tools and methods can help those in the organisational and institutional context to believe that progress on a project is going smoothly, even when in reality the project is experiencing significant problems

Conclusions

In this chapter, we have explored how and why projects are used to manage innovation processes. In doing this, we have looked at why the tools and methods that have been designed to manage projects are not consistent with the emergent nature of innovation, often creating problems for those involved. We have also examined the problems of sharing what is learnt in projects with others outside the project itself. Finally, we have considered how these issues associated with project management are more problematic in contexts of complexity, where the underlying science, technology or regulatory environment is still evolving. Nevertheless, we conclude that project management methodologies don't simply play a functional role; their continued use in complex innovation contexts is as much about the symbolic role such artefacts play as it is about ensuring progress continues in a planned and controlled way.

? DISCUSSION QUESTIONS

The following discussion questions are relevant to using this chapter in teaching exercises and discussions or for revision:

1. Why do organisations often use projects to manage their digital innovation processes?

2. **What is meant by a learning boundary? Is a learning boundary inevitable in a digital innovation project context?**

3. **Why is project management more difficult in complex innovation ecologies? Why might tools and techniques that attempt to plan and control project activities still be used, even in contexts where such plans must be constantly 'torn up and redrawn'?**

Case questions

The following case questions might also be relevant to using this chapter in teaching exercises and discussions:

1. Describe the type of project that is being undertaken in SkinTech. Is it complicated or complex? What leads you to this decision? Is there an example of digital innovation that you think has the same complexities?

2. Why were they finding it difficult to stick to project plans in each project? Identify the different sources of uncertainty that are present in a digital innovation project.

3. What were the consequences of trying to 'squeeze' things into project plans?

4. What does this case tell you about existing project management tools and methods in the context of complex digital projects?

Additional suggested readings

Edmondson, A. C. 2012. "Teamwork on the Fly," *Harvard Business Review* (90:4), pp. 72–80.

Rice, M. P., OConnor, G. C., and Pierantozzi, R. 2008. "Implementing a Learning Plan to Counter Project Uncertainty," *MIT Sloan Management Review* (49:2), p. 54.

References

Andriopoulos, C., and Lewis, M. W. 2009. "Exploitation-Exploration Tensions and Organizational Ambidexterity: Managing Paradoxes of Innovation," *Organization Science* (20:4), pp. 696–717.

Besner, C., and Hobbs, B. 2004. "An Empirical Investigation of Project Management Practice: In Reality, Which Tools Do Practitioners Use?," in *Innovations: Project Management Research.*, D.P. Slevin, D.I. Cleland

and J.K. Pinto (eds.). Newton Square, PA: Project Management Institute, pp. 337–351.

Bessant, J., Von Stamm, B., and Moeslein, K. M. 2011. "Selection Strategies for Discontinuous Innovation," *International Journal of Technology Management* (55:1/2), pp. 156–170.

Chesbrough, H. W. 2003. "The Era of Open Innovation," *MIT Sloan Management Review* (44:3), pp. 35–41.

Cicmil, S., and Hodgson, D. 2006. *Making Projects Critical: An Introduction.* New York: Palgrave MacMillan.

Desouza, K. 2006. *Agile Information Systems: Conceptualization, Construction, and Management.* Oxford, UK: Elsevier.

Desouza, K. C., and Evaristo, J. R. 2004. "Managing Knowledge in Distributed Projects," *Communications of the ACM* (47:4), pp. 87–91.

Dougherty, D., and Dunne, D. D. 2012. "Digital Science and Knowledge Boundaries in Complex Innovation," *Organization Science* (23:5), pp. 1467–1484.

Edmondson, A. C. 2012. "Teamwork on the Fly," *Harvard Business Review* (90:4), pp. 72–80.

Engwall, M. 2003. "No Project Is an Island: Linking Projects to History and Context," *Research Policy* (32:5), pp. 789–808.

Feldman, M. S., and Pentland, B. T. 2003. "Reconceptualizing Organizational Routines as a Source of Flexibility and Change," *Administrative Science Quarterly* (48:1), pp. 94–118.

Gann, D. M., and Salter, A. J. 2000. "Innovation in Project-Based, Service-Enhanced Firms: The Construction of Complex Products and Systems," *Research Policy* (29:7), pp. 955–972.

Gherardi, S. 2011. "Organizational Learning: The Sociology of Practice," in *The Blackwell Handbook of Organizational Learning and Knowledge Management,* M. Easterby-Smith and M.A. Lyles (eds.). Oxford, Melbourne, Berlin and Malden, MA: Blackwell, pp. 43–65.

Grabher, G. 2002. "The Project Ecology of Advertising: Tasks, Talents and Teams," *Regional Studies* (36:3), pp. 245–262.

Lindgren, M., Packendorff, J., Hodgson, D., and Cicmil, S. 2006. "Projects and Prisons," in *Making Projects Critical,* D. Hodgson and S. Cicmil (eds.). New York: Palgrave MacMillan, pp. 111–131.

Meredith, J. R., and Mantel, S. J. 2011. *Project Management: A Managerial Approach.* Hoboken, NJ: John Wiley & Sons.

Meyer, J. W., and Rowan, B. 1977. "Institutionalized Organizations: Formal Structure as Myth and Ceremony," *American Journal of Sociology* (83:2), pp. 340–363.

Nelson, R. R., and Winter, S. G. 1982. *An Evolutionary Theory of Economic Change.* Cambridge, MA: Belknap.

Newell, S., Goussevskaia, A., Swan, J., Bresnen, M. (2008) "Managing interdependencies in complex project ecologies: The case of biomedical innovation," *Long Range Planning* (41: 1), pp. 33–54

Newell, S., Goussevskaia, A., Swan, J., Bresnen, M., and Obembe, A. 2008. "Interdependencies in Complex Project Ecologies: The Case of Biomedical Innovation," *Long Range Planning* (41:1), pp. 33–54.

Orlikowski, W. J. 2006. "Material Knowing: The Scaffolding of Human Knowledgeability," *European Journal of Information Systems* (15:5), p. 460.

Pentland, B. T., Hærem, T., and Hillison, D. 2011. "The (N) Ever-Changing World: Stability and Change in Organizational Routines," **Organization Science** (22:6), pp. 1369–1383.

Pickering, A. 1995. *The Mangle of Practice*. Chicago: University of Chicago Press.

PMBOK. 2004. *A Guide to the Project Management Body of Knowledge*. Project Management Institute. Available at: http://www.pmi.org/.

Rice, M. P., OConnor, G. C., and Pierantozzi, R. 2008. "Implementing a Learning Plan to Counter Project Uncertainty," *MIT **Sloan** Management Review* (49:2), p. 54.

Sheremata, W. A. 2000. "Centrifugal and Centripetal Forces in Radical New Product Development under Time Pressure," *Academy of Management Review* (25:2), pp. 389–408.

Styhre, A. 2006. "Science-Based Innovation as Systematic Risk-Taking: The Case of New Drug Development," *European Journal of Innovation Management* (9:3), pp. 300–311.

Swan, J., Bresnen, M., Robertson, M., Newell, S., and Dopson, S. 2010. "When Policy Meets Practice: Colliding Logics and the Challenges of 'Mode 2'initiatives in the Translation of Academic Knowledge," *Organization Studies* (31:9–10), pp. 1311–1340.

Swan, J., Goussevskaia, A., Newell, S., Robertson, M., Bresnen, M., and Obembe, A. 2007. "Modes of Organizing Biomedical Innovation in the Uk and Us and the Role of Integrative and Relational Capabilities," *Research Policy* (36:4), pp. 529–547.

Swan, J., Robertson, M., and Newell, S. 2016. "Dynamic in-Capabilities: The Paradox of Routines in the Ecology of Complex Innovation," in *The Annual Series Perspectives on Process Organization Studies (P-Pros)*, J. Howard-Grenville, C. Rerup, A. Langley and H. Tsoukas (eds.). Oxford: Oxford University Press.

The Guardian. 2012. *Thalidomide scandal: 60-year timeline*. The Guardian, Available at: https://www.theguardian.com/society/2012/sep/01/thalidomide-scandal-timeline.

Thomas, J. 2006. "Problematising Project Management," in *Making Projects Critical*, D. Hodgson and S. Cicmil (eds.). New York: Palgrave MacMillan, pp. 90–107.

Townley, B. 2002. "The Role of Competing Rationalities in Institutional Change," *Academy of Management Journal* (45:1), pp. 163–179.

Van Norman, G. 2016. "Drugs, devices and the FDA: Part 1: An overview of approval processes for drugs." JACC: Basic to Translational Science, (1:3), pp. 170–179.

Wagner, E. L., and Newell, S. 2006. "Repairing Erp Producing Social Order to Create a Working Information System," *Journal of Applied Behavioral Science* (42:1), pp. 40–57.

Wood, M. 2002. "Mind the Gap? A Processual Reconsideration of Organizational Knowledge," *Organization* (9:1), pp. 151–171.

6 PROJECT LIMINALITY AND OPEN INNOVATION

Summary

In this chapter, we look in some detail at how innovation projects can be managed whether undertaken within an organisation's boundaries or as a more open process, typically seeking to expand innovation processes to include others outside of the organisation. In terms of internal projects, the concept of liminality is used to explore the trade-off between being separated from the ongoing workings of an organisation (and so less constrained by existing structures, rules and norms) and having to convince the rest of the organisation that the outcome is a good one at the end of the process. This highlights how the existing organisation can constrain innovation projects. Moving to a more open innovation process can reduce these constraints and potentially bring to the process more novel knowledge that can allow more creativity. Open innovation includes both inside-out and outside-in processes and can be applied to service innovation as well as product innovation. We need to recognise, however, that there are different types of open innovation and that the choice of open innovation process needs to be carefully considered. Decisions involve both how to search for open innovation partners and how to manage or govern the partners once selected.

Learning Objectives

The learning objectives for this chapter are to:

1. Consider the advantages and disadvantages of making an internal innovation project more or less separate (liminal) from the rest of the organisation

2. Explore why open innovation processes have been more widely adopted in the twenty-first century

3. Examine different types of open innovation and understand when these may be more or less appropriate

4. Review different 'search' and management problems associated with open innovation and their potential solutions

5. Understand the potential strategic significance of project liminality and open innovation.

Case: Project liminality and open innovation – Defence-co's InnovationJam and the city of Vienna's co-creation project

This case explores innovation initiatives in two different organisations. The first initiative examines a small-scale innovation project at a large national defence organisation in the UK (anonymised as Defence-co). Defence-co was looking to innovate to streamline processes internally within its departments and find ways of reducing bureaucracy by cutting organisational red tape. This innovation project was facilitated by an external partner, IBM, as part of their InnovationJam programme (sometimes referred to as 'Jamming'). The second initiative focuses on large-scale open innovation in a European city, Vienna, and how its tourist board used open innovation to help gain new knowledge and ideas to shape the future strategic direction of the city. Overall, the case is useful for understanding key elements of project liminality and open innovation.

Initiative #1: Defence-co's InnovationJam

Defence-co recruited IBM as consultants to help facilitate one of its projects through IBM's InnovationJam programme. InnovationJams or 'Jamming' has been highlighted as being an illustrative example of innovation through internal or external inclusion of a wider range of actors. IBM first used the term to describe its internal online conferences (Bjelland and Wood, 2008). Taking the name from the jamming activity of musicians, IBM set out to replicate the notion of creative collaboration between people who might never have met before (Bhalla, 2010). Jamming activity has also been linked in similarity to the concept of crowdsourcing (Howe, 2009) to capture 'the wisdom of the crowd' for organisations to explore and exploit new strategic directions. While InnovationJams were originally devised to help IBM innovate internally, they have since been marketed externally and generate significant revenue for the organisation as a consultancy tool.

IBM facilitated the Defence-co InnovationJam using IBM Connections, a web 2.0 platform. The jamming session lasted for two days and involved 67 participants, generating 90 innovation ideas with a combined total of 287 discussion posts. It was focused around three main topics as part of the project to improve processes and reduce bureaucracy at Defence-co. Figure 6.1 below shows a poster that was used by IBM and Defence-co to promote the InnovationJam and offers an overview of what a Jam is and how it works.

The event was formally arranged, involving nine months of planning and included using the above poster to promote the initiative and hosting a trial 'Jam' with IBM. Additionally, the activity was focused on the output from a previous initiative where the organisation had asked employees how they could cut organisational red tape in order to make their roles less restricted. It also had the aim of engaging employees in ongoing transformation of the organisation led by their new Chief Information Officer (CIO). An IBM executive responsible for organising the InnovationJam initiative stated:

> A new 3-star general came to the organisation and saw previous failings, which caused the old organisation to essentially fail and be rebranded. He came out and asked his employees, 'what needs to change', 'what's stopping the organisation from working'. From the red tape output, we asked if he'd be interested in doing a Jam for his employees ... and have a focused discussion.

An employee of IBM's Jam team emphasised that they work as formally arranged spaces that help separate this form of organising for innovation from the wider organisational structure. The interviewee's summary of Jamming also emphasised how the outcomes should be reflected afterwards in an attempt to integrate them into the wider organisation:

> We view a Jam as a short-term intervention where you can bring everyone together on a really focused set of topics, of significant consequence to the organisation. They are structured in such a way that the process we have behind it, it's not really the tools, it's very much the process. How you get engagement before the event with the key stakeholders, how they are aligned to the key issues that you're trying to challenge, and how you steer that debate over say three days, to have tangible outcomes at the end of the period with known owners and drivers and true

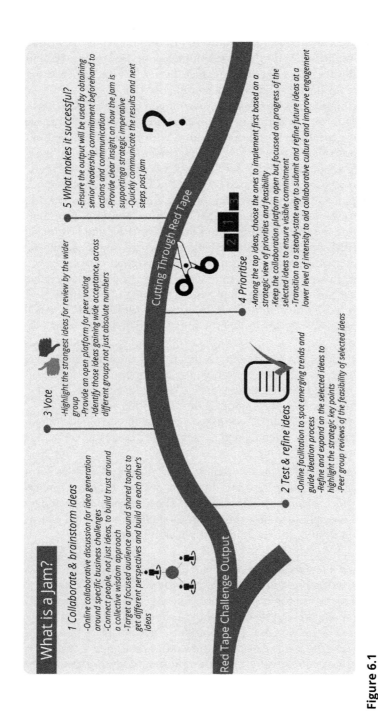

Figure 6.1

What is a Jam? Poster used to advertise the Defence-co InnovationJam (poster reproduced with permission from designer)

> engagement across the organisation. What I find with the Jams is that when a new CEO joins the organisation, they have their own views, their own strategy and it's an amazingly good way of getting the message to everyone in the organisation.

A member of the IBM Chief Technology Officer (CTO) team and moderator for the InnovationJam also confirmed that Defence-co was planning to use outcomes from the Jam, but it would need to consider carefully how it would integrate them into the wider organisation. This is confirmed in the Jam outcome analysis, where the organisation mentioned wanting to take forward at least three core ideas from each of the three Jam topics:

> We take a step back from the Jam once it's finished ... in this case it was positive, and I know they're working on inputting some of the ideas participants came up with. Exactly how I wouldn't be too sure, but it's something we'll keep an eye out for.

Ultimately, the InnovationJam was used by Defence-co to generate new knowledge and ideas around organisational innovation.

Initiative #2: The city of Vienna's co-creation project

The city of Vienna's co-creation project offers an example of open innovation and one which is significantly longer than that of Defence-co and structured in three main stages. The first was an 'open strategy meeting' with around 30 participants. These were mostly 'internal' stakeholders to the city, such as tourism executives and policy-makers but also people external to the organisation who were involved with local tourism, such as business owners. One of the organisers explained that this was a way of discussing strengths, weaknesses, threats and opportunities and a way of launching the open innovation process. Figure 6.2 shows the process in its entirety, from the initial meeting through stages of open innovation.

They further explained that the second phase, open innovation I, was one of two larger processes within the open innovation initiative, both of which utilised a web-based social collaboration platform. With the help of a consultancy firm, the Vienna tourist board implemented the technology platform and invited over 650,000 people from around the world to contribute to an idea contest about how the city might improve its tourism. This lasted one month, involved 800 participants and generated 546 ideas. The 'best' ideas from this were refined for the final stage, open innovation II, and focused into eight main topics; around 2,500 participants from the city were invited. These included policy-makers, business owners and other stakeholders who participated openly using the same online platform. Of this number, 255

Figure 6.2

Overview of the city of Vienna's co-creation project

contributed to further discussion and refinement. One manager from the Vienna tourist board described this phase as being a more focused part of the initiative, with the aim of getting buy-in for ideas:

> So, this was really important because they all had the possibility to comment on ideas. There were 237 idea comments, and 74 buy-ins, where somebody said they will engage and help with this idea. The target groups within the second phase were the key stakeholders, politicians, tourism experts working in the (city name) tourism industry. Our advisory board rated them also, the best 10 or 20, and these were included in the final strategy.

The final reflections from the organising team were that, although the project was seen as a success, winning awards for its innovative format and resulting in new innovation for the city and also a five-year strategic plan, some of the structures used have not successfully translated into the routines of the organisation as much as they would have hoped:

> I think there would have been the possibility to do a bit more, and to have more results and more buy-ins, and to do more with the results from this open process, and to keep the engagement level quite high. How do you keep this alive? how do you carry forward ideas? I think this is a point where we may not have succeeded. We could have done a bit better.

In this case, the organisation was able to suspend conventional innovation and strategy approaches through an open, digitally enabled initiative. However, although now informed by a wealth of insights, managers decided not to persist with further open innovation initiatives. The initiative saw substantial adaptation to innovation and also to strategy, but this was not sustained to feed into transforming long-term practice in the organisation, also emphasising the potential temporality of innovation processes in organisations.

The two initiatives discussed here are used to help illustrate and discuss key topics and themes, as we expand on project liminality and open innovation in this chapter.

Introduction

As we saw in Chapter 5, and in the case of Defence-co's InnovationJam, projects are an important organisational route for developing innovative products and services as well as new organisational arrangements. The key idea behind projects as a form of organising is that they provide a degree of separation from the ongoing workings of the organisation, which can allow those involved to access new ideas and think differently. We build on the ideas introduced in the case in this chapter and discuss two aspects to this separation between the project and the ongoing organisation. First, we explore the concept of liminality as a way of thinking about how and why a project team's separation from the ongoing organisation can promote more novel ways of thinking but also how this might create problems for the rest of the organisation accepting what the project team comes up with. This allows us to think about the degree of liminality likely to be most sensible in any given innovation context. Second, we look at the more radical solution, which is to abandon the idea that innovation is something that an organisation does itself and rather to move towards an open innovation model. As was illustrated with the city of Vienna's co-creation project, partnerships are used in open innovation, and the organisation recognises that new ideas and knowledge can come from elsewhere, not just from within its own boundaries, however much space the project team are given to avoid being restricted by current organisational ways of doing things.

Project organising and liminality

In Chapter 5, we discussed the idea of a learning boundary. Another way of thinking about this learning boundary is through the concept of liminality. A *liminal space* is one where individuals can generate new ideas and practices free from the constraints of ongoing organisational routines (Hendry and Seidl, 2003). 'Liminal' is derived from the Latin word *limen*, which means threshold. The term was used by van Gennep (1909) when he was writing about rites of passage. He identified three phases as a person moves from one social status to another – *separation* (from the old social status), through a *transition* phase, to emergence through *incorporation* (into the new social status). In some societies, the transition phase is very pronounced (e.g., the Maasai people in Kenya send adolescent boys away from the village to live together for anything from 8 to 12 years to learn to become warriors). In many Western countries, the transition to adulthood is fuzzier – teenagers gradually get the right to marry and to leave school (16 in the UK and USA, for example), to drink alcohol (18 in the UK, and 21 in the USA) and vote (18 in both). By contrast, Article 3 of the Japanese Civil Code states that the age of majority is 20 years of age whereas the age of consent is 13 under the Japanese national criminal law code.

In many Western and Asian societies, then, there are a series of separations, transitions and incorporation phases so that the individual gradually shifts from childhood to adulthood. For the Maasai, the transitory state is for a significant period, during which the adolescents are clearly separated from the rest of their village and their return to the village is a single passage point back to incorporation. The latter more isolating transitory state provides more opportunity for radical change – the boy who left the village is very different from the man who returns. However, this can also make re-entry difficult precisely because so much change has occurred on the part of the individual, whereas village life itself may have changed very little. Projects likewise can be described as liminal and also exhibit a more dramatic or gradual transition from and to the rest of the organisation. For example, in some innovation projects, those involved are placed on the project for a period of time, with their ongoing jobs back-filled so that they can focus exclusively on the innovation project. In this type of project, they are often given a physical space which is separated from the rest of the organisation. In other projects, those involved work part-time on the project but also continue to work in their functional home. Project meetings on these projects will typically be in a very local office space because those involved need to move back-and-forth between the project work and their ongoing job. And just like the example of adolescence, we can predict that the more separated is the project team from the ongoing organisation, the more different may be the solutions that are produced, but also the more difficult will these solutions be to communicate to the rest of the organisation as the acceptable way forward. In the Defence-co case, we saw that separating staff members into a focused innovation initiative (the InnovationJam facilitated by IBM) created a liminal space which encouraged ideas and knowledge from outside usual organisational routines. The ideas here seemed to be quite highly thought of by senior management; however, the challenge that remained was how to integrate these into the wider organisation. Similarly, the TechCo case in Chapter 2 provides an example where the project team rolled out a digital system that users did not understand and did not initially find useful. In this case, the project team was separated (i.e., very liminal) from the rest of the organisation for the duration of the project, with their roles back-filled. The Kodak digital camera project is also an example where the team that worked on this innovation were separated in order to develop a very innovative solution compared to the film-based technology produced by the company up until that time. In this case, again, the solution the innovation project arrived at was not accepted by many decision-makers in the organisation.

Liminality as a concept in innovation

We can explore the concept of liminality in more detail to understand both the increased creativity potential of more liminal projects but also the sources of frustration for such projects. During the transitory or liminal phase, those involved are 'betwixt and between the positions assigned and arrayed by law, custom, convention and ceremonial' (Turner, 1969, p. 95), living 'at the limits of existing structures' (Tempest and Starkey, 2004, p. 507). The transition phase is therefore liminal in the sense that it is a period of ambiguity, with those involved being in a social and cultural limbo. This limbo state has both negative and positive connotations, and people in a liminal phase are 'temporarily undefined, beyond the normative social structure' (Turner, 1982, p. 27; Hendry and Seidl, 2003). Therefore, in many societies, adolescence can be described as a liminal phase, because a 'young adult' is one who does not yet have the full rights of an adult but is also no longer a child with limited responsibilities. While this 'weakens' young adults, since they do not have certain rights over others, it also 'liberates' them from structural obligations. In this sense, liminality can be both painful and enjoyable simultaneously (Rottenburg, 2000), freeing the person from institutional constraints but at the same time marginalising them. A project team likewise can be liberated from organisational constraints but marginalised from the ongoing decision-making within the main organisation.

The term liminal has been used to describe the condition of temporary employees in flexible organisations (Garsten, 1999), to discuss the consulting experience (Czarniawska and Mazza, 2003) and to consider individual and organisational learning experiences (Tempest and Starkey, 2004). It can also be used to describe the project experience (Wagner et al., 2012). We consider this next as we look at the various characteristics of a liminal space (Figure 6.3).

Characteristics of liminal space

One of the first characteristics of a liminal phase is that it is *temporary*, hence Garsten's (1999) use of the concept to explore the experience of temporary workers. In the Defence-co initiative, we saw this temporality with the InnovationJam lasting just three days (though taking a significant amount of time to plan). Organisations use temporary employees precisely because this allows them to be more flexible – adding and reducing headcount as and when needed – and, in doing this, creating 'liminal subjects in flexible organisations' (Garsten, 1999). Project teams are very similar since they are also used to give organisations more flexibility – to be involved in some kind of innovation process while business-as-usual can continue in the rest of the organisation. A second characteristic of a liminal phase is that it is *ambiguous*, as people in liminal space 'slip through the network of classifications that normally locate states and positions in cultural space' (Garsten, 1999,

Figure 6.3

Liminality represents a 'passage' from one condition/state to another
©*Getty Images*

p. 606), presenting both risks and opportunities simultaneously – for both
the individuals and the organisations involved (Tempest and Starkey, 2004).

For Turner (1982), this ambiguity provides those in a liminal phase with
the opportunity to upset normative orders and transcend institutional
boundaries: an opportunity for creativity. The ambiguity exists because activities are performed 'backstage' – separated from mainstream organisational
activities. This enables those involved in a project to think and act differently
and so to envision arrangements that are not constrained by existing work
practices and routines. This is related to the third characteristic of a liminal
space, which involves *freedom*. Those in the liminal space are freed from
structural and institutional constraints and obligations (Hendry and Seidl,
2003). An organisational structure entails systems of classifications, models
of thinking and norms governing what is considered acceptable. A project in
a liminal space can challenge this structure and its associated norms and
systems of meaning (Garsten, 1999). This aspect of liminality provides freedom in two senses: freedom from institutional obligations (technical and
bureaucratic) prescribed by the organisation but also freedom to transcend
existing ways of thinking and norms and so try out new things (Turner,
1982). Indeed, for Defence-co staff, the InnovationJam effectively flattened
the organisational hierarchy temporarily, and all staff involved were treated
equally in their idea and knowledge generation. Likewise, Hagen et al.

(2003) discuss how the liminal space provided by education (specifically studying an MBA) can provide the opportunity to critically examine dominant orthodoxies. Ultimately, during the liminal phase, creativity and innovation can be enhanced since innovation occurs most often at interfaces and thresholds.

A final characteristic of a liminal phase is that it encourages a strong *sense of community* with those sharing the experience. Simultaneously, however, it can create a divide with those 'on the outside' (Czarniawska and Mazza, 2003). This strong sense of community within the project team is important because, as Wenger (1998) argues, knowledge generation rests on sensemaking, and sense-making occurs within the context of *communities of practice* (Lave and Wenger, 1991). The paradox here is that there may be a strong community among those sharing the liminal space (i.e., the project team), while the lack of connection with those within the rest of the organisation can mean that liminality threatens relationships with others across the organisation (Tempest and Starkey, 2004). In this light, Garsten (1999) describes how temporary employees have limited access to information channels and locales. More importantly, applying these ideas to project contexts, the knowledge generation and sense-making that have occurred among those in the project team are not extended to the rest of the organisation, who will therefore remain rooted in the old practice routines. This sets the scene for resistance at project rollout (as with TechCo in Chapter 2), when the new practices produced by the project team rub up against those still wedded to old arrangements. This was also probably an issue for Defence-co, even though some ideas and knowledge generated were positively received (Figure 6.4).

Paradoxes in projects and liminality

Thinking back to the earlier ideas about social status in relation to liminality, and the difference between Maasai and European teenage liminality, we can identify that ensuring project participants maintain connections and communications with the rest of the organisation – for example, not having project team members assigned full time to a project – can increase the chances that the ideas and technologies created by the project team will be better accepted. At the same time, the ideas and technologies created may be less radical because the team has not been completely isolated from existing organisational pressures and norms. In other words, there may be some tradeoffs in terms of how isolated (and so liminal) to make a project team tasked with creating an innovative product, service or new organisational arrangement. Indeed, potentially the degree of liminality can be selected based on how innovative the solution needs to be. In the Defence-co case, the degree of liminality was quite high but very short-lived in its temporal nature.

Figure 6.4

In projects, a sense of community within a liminal space might be detrimental to those outside a community

©PhotoDisc/Getty Images

Key Concepts: Liminal Characteristics of Projects

Temporality:
Liminality is, by definition, only temporary. A project is a temporary assignment so that the team member knows that they will eventually move on, although it may be unclear what this move will involve

Ambiguity:
A liminal space is ambiguous in the sense that a person does not fully understand what is involved. In the context of a project, this liminal characteristic can cause some stress, especially in the early stages, but can also be the opportunity for creativity

Freedom to/from:
In a liminal condition, a person is free to experiment because they are less constrained by the norms and expectations of the social setting. In a project, this provides the opportunity for experimentation with new ideas to see what might work

Community:
This characteristic of the liminal phase encourages both a strong sense of community in the project team while possibly creating a divide with those on the 'outside', that is, people who have not been involved in the project in the rest of the organisation

This discussion of liminality, as it may apply to project work – and so innovation – recognises its paradoxical nature, having both positive and negative implications. These are positive in that play and transcendence of existing institutional structures provide the opportunity for those involved to think creatively and come up with transformational ideas and practices. Equally, they are negative in terms of marginality and exclusion among those not involved and also the potential stress for project team members in relation to both the ambiguity of what they are doing in the project but also what they might be asked to do once the project is finished. Moreover, the move from the transition phase to incorporation is crucial. It is likely that where the transitory liminal phase is extreme (i.e., there is a prolonged period of separation as with Maasai boys and an isolated project team), there is more opportunity for a radical or transformatory solution to be generated because the project team's thinking can be less constrained by current systems and practices. This may set the stage for a more difficult incorporation because it creates the potential for a tension when the project team presents its innovative ideas and knowledge.

How the project team handles this tension then becomes crucial. Wagner et al. (2012) examine how a language of peace-making (attempting to communicate with, understand and accommodate the 'other side', often involving compromise, in order to get an agreement that all can accept) can be helpful for resolving tension and encouraging connectivity between the learning in the project and the subsequent learning of users in the wider organisation. This is well illustrated by the TechCo case in Chapter 2, where 'peace-making' was absent during the first rollout of their new Enterprise System. Indeed, in this case, there was no attempt to engage users at all; rather, users were told simply to 'use the new system', with the result that they simply ignored the system and carried on using their legacy systems. The second rollout ran more smoothly precisely because those leading the project were more careful to listen to users and accept that they would influence the final deliverable system.

This analysis of the project experience focuses on why projects need autonomy so that they are not constrained by ongoing organisational routines. However, even when the project team is provided with considerable separation and autonomy, there is still the issue that the project is led by the organisation, with most people coming from within the organisation (albeit often released from their ongoing work routines) even when some external partners may be involved (e.g., IBM consultants used in the Defence-co InnovationJam). A more radical solution is not to assume that the organisation needs to lead the project itself but instead sees that big problem 'chunks' of an innovation project, or even the entire project, can be 'outsourced' to third parties. We discuss this next as we consider in more detail open innovation processes, a phenomenon already touched upon in earlier chapters. The city of Vienna's co-creation project is used here to illustrate key ideas and themes.

Open innovation

Open innovation was heralded as a new paradigm for organising innovation when it was first presented by Chesbrough (2003) in the book *Open Innovation: The New Imperative for Creating and Profiting from Technology*. The 'old' model of innovation, according to Chesbrough, was based on 'self-reliance' or DIY (do-it-yourself). That is, firms were successful to the extent that they invested more heavily in R&D than their competitors, with the result that they were able to bring more and better new products and services to market more quickly. Firms would protect their intellectual property (e.g., via patents) produced through this R&D effort in order to ensure that they, and only they, could further exploit it (Porter 2004). However, open innovation turns this approach on its head, with firms actively commercialising external (not simply internal) ideas and by using outside (as well as in-house) pathways to the market. The basic premise of open innovation is quite simply that 'most smart people work for someone else' so that knowledge is widely distributed. If a firm can tap into this distributed knowledge rather than rely exclusively or even mostly on internal knowledge, then it is likely to come up with more and better ideas for new products and services. As we saw with the city of Vienna's co-creation project, open innovation principles, such as tapping into external sources of knowledge for new ideas, were the central driver of the initiative. This, coupled with the increasing ability to be able to tap this distributed knowledge through internet-enabled (i.e., digital) communications, such as through web-based collaboration platforms, heralded the way for this change in understanding of innovation processes.

The differences between 'closed' innovation and open innovation models of innovation are summarised in Table 6.1.

The differences suggested between these closed and open innovation models appear to be stark, yet even when innovation is approached in a more closed way, organisations still acquire ideas and resources from the external environment. As we discussed in Chapter 2, an organisation's absorptive capacity – its ability to recognise, assimilate and exploit external sources of knowledge – is important in relation to innovation (Cohen and Levinthal, 1990). Nevertheless, as we identified, absorptive capacity is recognised to be path-dependent – an organisation will recognise the value of external knowledge to the extent that its employees have related knowledge that allows them to understand and evaluate pertinent new external knowledge. Knowledge that is very different from what is already known will tend to be ignored, meaning that there can be 'lock-in' to a particular technological trajectory that may become obsolete as disruptive technologies shake up an industry (as discussed previously in the case of Kodak where the company was locked into its model of making profits through selling film for cameras and tried to emulate this in a digital photography environment where this business model was more difficult to use). This is where working in collaborative alliances and partnerships (i.e., networks supporting open innovation) can be advantageous. The city of Vienna saw the advantage of this in opening its open

Table 6.1 Differing assumptions of closed and open innovation approaches

Closed innovation assumptions	Open innovation assumptions
• Gather cleverest people as employees	• Recognise knowledge and expertise beyond firm boundaries
• Discovery, development and commercialisation must all happen internally	• Some parts of the discovery, development and commercialisation process can be done outside and still allow the focal firm to profit
• Will be only one winner – firm that can commercialise first	• Can be multiple winners collaborating together
• First to market will get the profits	• Often business model is as important as time to market
• Need to hoard intellectual property (IP) from competitors so firm can protect and profit from this	• Since a firm can benefit from others' IP, the firm can also share its own IP and collaborate with others
• A firm needs to have the best ideas internally to 'win'	• Winners will be those who can make use of best internal and external ideas

innovation co-creation project to key stakeholders in the city, such as policy-makers, business owners and inhabitants, and indeed outside the city, such as tourists. Not only do such partners have different knowledge so that a broader range of knowledge can be absorbed, partners can also be more easily changed than in-firm resources. This means that open innovation can lead to more disruptive innovation. However, we should not forget that focusing on, and protecting, core capabilities can be a source of competitive advantage (Gupta et al., 2009) and that knowledge-sharing is not 'risk-free' (Marabelli and Newell, 2012; Trkman and Desouza, 2012). This suggests that a firm will need to carefully decide what and when to engage in open innovation versus managing the innovation process in house (Figure 6.5).

Who is involved in open innovation?

The different parties that may be involved in an open innovation process are very varied and include customers, users, suppliers, universities and even competitors. The key idea behind open innovation is value co-creation – that is, value is not created by a focal firm and then exploited for its own benefits (e.g., in terms of profits) but rather the value is created collaboratively by different stakeholders who each benefit from being involved in the process.

 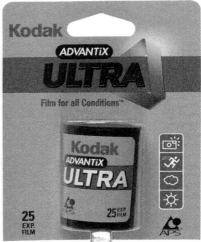

Figure 6.5

Kodak was not able to understand long-term challenges of being innovative
©MACMILLAN

How these different parties are involved in the innovation process can be very varied, including contests and tournaments (as demonstrated with the city of Vienna's web-based idea contest platform), alliances and joint ventures, licensing and open-source platforms. Felin and Zenger (2014) define four different types of open innovation arrangements:

1. *Markets and contracts*: in this type of open innovation, the focal firm contracts with another firm to provide a solution, which may be the entire innovation it is attempting to develop or a part of it. The assumption is that the supplying firm has more knowledge than the focal firm and so can develop the solution better than the focal firm could do itself. The knowledge transfer in this type of open innovation is relatively limited, with the supplying firm providing the complete solution rather than the knowledge that has enabled it to create this solution.

2. *Partnerships and alliances*: in this type of open innovation, the focal firm works collaboratively with other organisations jointly to solve problems and come up with new solutions in the innovation space. Since those involved are working together, there is going to be more knowledge transfer between the focal firm and the partners and alliances that it is working with.

3. *Contests and platforms*: in this type of open innovation, the organisation truly 'opens up' to a wide range of individuals and organisations that may possess knowledge relevant to solving a problem or providing a solution

related to an innovation process that it is working on. A digital platform is often used to allow participation from people who might be widely distributed. The idea of contests is to access knowledge from very disparate sources that might provide solutions that the focal firm would never have been able to think of itself. It is very closely associated with the city of Vienna's co-creation project introduced in this chapter and also with crowdsourcing, which is discussed in Chapter 8. Since the idea of the platform is to broadcast the problem to a wide range of people, the focal firm does not need to identify where relevant knowledge may be – it is up to those who chose to participate to decide whether they have knowledge that could be useful in solving the problem. More generally, the idea of a contest is that someone might participate who brings a very different perspective to the problem domain so that radical solutions are identified. The focal firm's main task is to provide motivation for people or firms to participate. The motivation may be some kind of prize, but there is also often intrinsic motivation for participation, with those deciding to do this finding the experience rewarding since they are challenging themselves to find a solution. In the city of Vienna initiative, the reward was that participants were able to see 'buy-in' for their idea and witness their idea come to life. Participants may also gain reputational benefits, as has been well documented in relation to open-source software development. This approach to open innovation, however, works well only where the problems to be worked on can be clearly specified in advance, meaning that the problem needs to be highly decomposable.

4. *Users and communities*: the final example of open innovation relates to situations where user communities develop their own solutions to their own problems, as with open-source software. Today, some firms are attempting to exploit the knowledge of their own users by setting up firm-hosted, digitally supported user communities that can contribute ideas for innovation. Users get involved in these community efforts because they hope that their contributions will lead to improvements in the product or service that they themselves use so that they will directly benefit from these improvements. Given the self-interest in users' involvement in these communities, the sharing of knowledge can be very open since they do not really have a vested interest in keeping knowledge to themselves (unlike in contests where giving away one's ideas may allow someone else to win the competition). An example in the city of Vienna case would be business owners and inhabitants potentially acting as a community and sharing knowledge to benefit the prosperity of the place where they live and work. A disadvantage, however, of this approach to open innovation is that the focal firm has little control over what the users decide to focus on in relation to a product or service improvement. We build on this again in Chapter 8 when we discuss social media innovations.

Inside-out and outside-in open innovation

Chesbrough (2003), in his original discussion of open innovation, made the point that any part of the innovation process (from the basic science in the laboratory right through to the marketing of the innovation) can be made open. He developed this idea by distinguishing between *outside-in* and *inside-out* open innovation. Outside-in tends to be the 'normal' focus of open innovation processes, where a focal company brings in third parties to support an innovation process, whether this is through a contract or a crowd-sourcing project, for example. Inside-out, however, can be just as important and relates to where a focal company decides to let go of some of its innovation projects and allow outside companies or individuals to exploit this in-house work.

It is perhaps strange that more companies do not develop an inside-out innovation process, given how many innovation projects started within a company are terminated before they reach the market. Sometimes, this is because the project was simply a bad idea that was never going to have market value, and, in these cases, it is probably good that a project was terminated before being launched on the market. Indeed, it is the case that projects can develop their own momentum which can reduce the likelihood that thorough risk analyses are undertaken, with the result that products or services are brought to market that should have been stopped well before they resulted in market failure. For example, Royer (2003) looks at 'why bad projects are so hard to kill'. She provides the example of RCA, which continued to invest in its SelectaVision project (to produce a videodisk player) long after all of its competitors had abandoned this technology, recognising how the improvement in VCR technology would make videodisks obsolete. In total, RCA invested USD $580 million in this technology over 14 years, with the product only on the market between 1981 and 1984 when it was finally abandoned in the face of poor sales. Rather than seeing this type of failure as being the result of managerial incompetence or bureaucratic inertia, Royer argues that the inability to kill doomed projects is often the result of an over-belief in the new product or service among the innovation/project champions and other key managers. Ironically, such belief is important to maintain interest and support for an innovation project over its life cycle of development, when there will be inevitable set-backs and problems that can reduce commitment. However, the dark side of this managerial belief in 'their project' is that a project can be kept going even when the evidence mounts that it is unlikely to be a success. We saw in the city of Vienna initiative that, indeed, although the co-creation project was seen as a success in generating ideas, senior management remained unconvinced of the value of including external sources of knowledge in future tourism strategy projects. Interestingly, the potential dark side of innovation led Royer to suggest that companies should have 'exit champions' as well as project champions. Exit champions are there to explicitly question the viability of a project through demanding hard data and should try to disrupt the prevailing beliefs of those who are very involved and committed to the project. Exit champions are thus there to make the case to kill a project even while a project champion might be making the case to increase resources allotted to that project.

Nevertheless, while there are clearly 'bad projects' in the innovation portfolio of many companies, it is also the case that there are potentially 'good projects' that are abandoned perhaps because of a lack of resources at a particular point in time or more often because it is decided that they don't fit the business plan or business model of the company. Chesbrough refers to these as false-negative projects – projects that were evaluated internally as not deserving continued investment but which might then also have the potential to be 'leaked' and turn out to be very valuable to another company. For example, Xerox invested in the development of the first practical personal computer at its Palo Alto Research Centre (PARC) but decided not to actually pursue this technology because it had decided that it was a copier company not a computer company. This was a strategy that at the time was seen to be important to defend its core copier business from competition, especially from Japan. Some of the technologies that were developed under this personal computer initiative at PARC included the use of icons, windows, point-and-click commands and local area networks – all technologies that were subsequently developed by other companies as personal computing took off.

Chesbrough suggests that the key to open innovation is for a firm to have a clear business model, which will help it make decisions about not only what knowledge to try to bring in from the outside (outside-in open innovation) but also what projects it will not be most sensible to continue to develop internally, even when these same projects might have some value to others if they are moved to the outside (inside-out open innovation). In this sense, we can define open innovation as a situation where an organisation purposively decides to include others from outside its own boundary, using financial and/or non-financial incentives to source all or part of its innovation portfolio that aligns with its business model.

Business models and innovation

The concept of business model then is key to Chesbrough's idea of open innovation. A business model relates to the question of how a company makes money by producing something (a product or service) that customers value. So, it requires that a company address the two questions that Peter Drucker many years ago identified as central to any organisation: Who is the customer? And what does the customer value? Importantly, organisations can and do innovate in their business models over time and, indeed, disruptive digital innovation is often associated with the introduction of a new

Figure 6.6

Business models are relevant to execute innovative strategies
©Getty Images/Cultura RF

business model into an existing competitive environment (as we discussed earlier). For example, Amazon (and other online digital retailers) have disrupted high-street retailers by introducing a new business model related to what customers value – the convenience of being able to make purchases online and have them delivered directly to their home rather than having to 'go to the shops' in the restricted times that shops are open. The fact that companies change their business model over time can account for why some innovation projects no longer fit their portfolio (as with Xerox), making inside-out open innovation attractive to reap at least some return from innovation investments that the company will not itself take to market. One practical tool to help here is the business model canvas, which is now commonly used by organisations (Osterwalder and Pigneur, 2010). The basis of the canvas is to enable organisations to develop a new, or document an existing, business model. The canvas is essentially a visual 'map' split into key sections which help describe an organisation's key activities, resources, partners, cost structure, value propositions, customer relationships, customer segments, channels and revenue streams. This is not only useful when considering Drucker's questions above but as an example emphasises that organisations need to adapt new and potentially collaborative, creative ways to analyse and innovate their business model over time (Figure 6.6).

Inside-out innovation can be particularly useful to consider not only when it is decided that an innovation project does not fit with the business model of the firm but also when there are resource constraints on innovation in lean times (Chesbrough and Garman, 2009). In a challenging business climate, it is sensible to focus on projects that are going to produce the most short-term gains and this can mean abandoning potentially fruitful but more risky projects with potentially longer-term pay-offs. If these projects are not to be simply abandoned, then the strategy of inside-out open innovation can be considered. For instance, rather than abandon a project, a company can decide to become a customer or supplier to a project that was previously being developed internally. This can be done, for example, by spinning out a new company as a separate entity that will take the innovation forward or by licensing IP to others who can further develop it. For example, Philips was a global leader in consumer electronics, but this industry faced very tough competition from Asian manufacturers. In the face of this competition, Philips decided to focus on healthcare and wellness markets and spun off its semiconductor business and simultaneously started to capitalise on its thousands of patents that had been developed for its legacy electronics industry but which were no longer relevant to its more focused business model. This allowed the company to earn millions of euros annually from licensing this IP. It has also innovated and tapped into the booming 'smart home' market with its Philips Hue smart lighting. Of course, as Chesbrough and Garman note, it is important in these inside-out processes to beware of pushing new technology out for others to develop who will then become competitors and so threaten the focal firm's existing markets. So, these are strategic decisions that must be made at the corporate level rather than the business unit level.

Another way that companies can encourage inside-out innovation is to set up incubators, sometimes with some investment from the focal organisation but also with the remit of attracting additional sources of funding for the new ventures. The new ventures supported in this kind of incubator will often be developing technologies or ideas that were initially developed within the focal company. However, these projects within the incubator are less risky to the focal company because they are developed independently of the company, although if successful they may be brought back inside the company. Again, however, it is important to recognise that ventures thus 'incubated' can become competitors.

Open innovation and services

Open innovation is often considered in relation to the development of new products. However, Chesbrough (2010) argues that the idea is also relevant in respect of services. This is important given that, as we have already seen in Chapter 1, digitally facilitated service innovation is increasingly important, even for product-based firms. Chesbrough conceives of outside-in and inside-out open innovation as being just as relevant to service offerings as product offerings. He uses the example of Lego to illustrate the outside-in service innovation potential. Lego encouraged user innovation by providing programmable motors with its plastic bricks so that customers could design new Lego creations that move. However, someone hacked the software so that the motors could do things that Lego had not programmed them to be able to do. Initially, Lego tried to stop this, viewing it as illegal. Then it decided to make the software intentionally open so that customers could reprogram the motors. Its decision was based on the idea that those outside the company might be able to find creative new opportunities for the Lego product. And sure enough, an external party developed a school curriculum to teach children about robotics using Lego. This, then, is a new service that Lego can offer related to digital education (Schlagwein and Bjorn-Andersen, 2014) (Figure 6.7).

Chesbrough's example of inside-out innovation is based on Amazon. Amazon was an early success story in selling products over the internet and gained a lot of very valuable knowledge about how to do this successfully by, for example, using customer reviews. As other retailers realised the opportunities of selling online, rather than in store, Amazon's knowledge of online retailing became increasingly valuable. Amazon could have decided that this was its proprietary knowledge and so have hoarded it. On the contrary, Amazon started a new service to help other retailers develop their online selling sites. Later, it offered its own infrastructure to these outside retailers, hosting their sites on its own servers and even performing the fulfilment

Figure 6.7

Lego's (open) innovative approach to services
©Getty Images

part of the transactions for these retailers. In this way, Amazon has been able to exploit its own knowledge by offering services to other online retailers and has reaped the rewards.

These examples show that being more open with respect to customers – realising that the customers are more likely to provide insights into what would be useful to them than are the company's own internal R&D and marketing units – can be a major source of innovation, especially around services. This is quite different, argues Chesbrough, from Michael Porter's traditional view of a value chain, which saw service as occurring at the last stage, after the product was designed, built, marketed and sold. Service in this value chain model is simply the part that ensures customers remain satisfied after they have purchased a company's product. In the open innovation service model, value is rather seen to be co-created by continuous interaction and knowledge-sharing between the focal firm and its customer(s). This sharing of knowledge, moreover, will include tacit knowledge, which can only be shared, as we have seen, by a more continuous interaction between the parties involved. UPS (United Parcel Service), for example, no longer just delivers parcels for its customers; for some organisations, the company takes over the whole shipping function, regardless of whether the parcels are delivered by UPS. This open innovation process allows UPS to understand more about its customers so that it can develop innovative solutions for new services.

Open innovation and networking: Brokering and boundary spanning

In relation to acquiring external knowledge for open innovation, we need to consider some important concepts from network theory. Here, the concept of *structural hole* is helpful. Burt (1992) used the term to describe the situation in which a person connects to two communities (or individuals) but these communities (or individuals) do not connect to each other. Ultimately, the person who *brokers* communities has access to information that others do not and can use this to their advantage – for example, to control resources and influence perceptions. In a sense, the middle managers in CommCo, whom we introduced through the case in Chapter 3, felt threatened because the structural holes that they had traditionally brokered were threatened by allowing users to talk to each other freely (through user-generated content (UGC) channels), bypassing the traditional hierarchical form of communication. This is why they resisted the implementation of the UGC channels. In this case, the structural hole was internal to the organisation. A structural hole can also be across different organisations and this is then more relevant to open innovation.

Those who broker across structural holes have been found to be more successful in their careers, although brokers can also become distrusted when parties see them exploiting their position for personal gain. Furthermore, brokers may be limited in what they can learn from the disconnected parties because the fact that they stand outside both communities means that they may not have access to all the valuable tacit knowledge (Obstfeld, 2005). This accounts for the fact that there have been contradictory findings about bro-kering roles and their advantages: not all studies have shown that brokers can gain benefits from their position across a structural hole. In examining this issue in more detail, Fleming and Waguespack, (2007) distinguish between brokers (those who bridge structural holes) and *boundary spanners*, those who bridge across communities by translating knowledge between the communi-ties so that each can better understand the other and share their knowledge. We saw this in the Defence-co case, where IBM consultants played a signifi-cant role in moderating on the IBM connections platform, pushing discus-sion forward to ensure that efficient and effective ideas and knowledge were generated. Boundary spanners often do not occupy a structural hole. For example, a boundary spanner may belong to one of the communities but have sufficient knowledge of the other to translate knowledge across the communi-ties. For instance, Earl and Skyrme (1992) discuss the importance of the 'hybrid' manager who can build a bridge between the IT department and other business departments because they have some understanding of both digital technology and business disciplines. Or a boundary spanner, involved in an open innovation project, may belong to one organisation but know a lot about another involved in an alliance (e.g., because perhaps this person used to work for that organisation) and so help to share knowledge across the part-ners. Fleming and Waguespack (2007) identify how both types of networking are important in relation to leadership positions, but make the point that boundary spanning between technological areas is more important for suc-cessful leadership than is brokerage, at least in open innovation communities. They explain that this is because of the importance of technological knowl-edge translation for open innovation processes – hence the requirement for boundary spanning – and because brokers can become distrusted (Figure 6.8).

The above suggests that it is not sufficient simply to be in a position to bridge a structural hole. Bridging a structural hole will expose the broker to diversity – diversity in terms of multiple partners with different goals and interests and diversity in terms of knowledge. The broker, then, needs to be able to manage this diversity, and this requires both process- and knowledge-based capabilities (Vasudeva et al., 2013). In the city of Vienna initiative, the tourism board split stakeholders based on their knowledge and interests. In Figure 6.9, stakeholders highlighted in red were involved in open innovation I – the idea contest welcoming ideas from a broad group of stakeholders such as tourists, inhabitants and students. Stakeholders highlighted in grey, how-ever, were involved in open innovation II – the key stakeholder discussion that focused on more refined discussion of ideas between the tourism indus-try, business owners and policy-makers.

Process-based capabilities refer to individuals managing partner relation-ships themselves and attempting to coordinate these partners even when

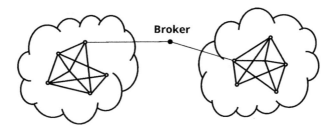

Figure 6.8

A diagram representing a broker between two communities

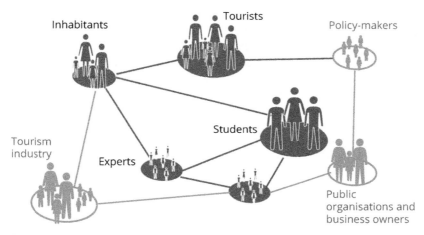

Figure 6.9

Overview of stakeholders and their involvement in the city of Vienna's co-creation project

there may be divergent or contradictory goals. This is likely to require considerable social skill. Knowledge-based capabilities refer to being able to integrate the diverse knowledge accessed through the partnership by, for example, acting as a boundary spanner and translating knowledge rather than simply acting as a knowledge broker.

Open innovation: The search issue

Above, we have recognised that open innovation success depends on both the identification of new sources of knowledge and the ability to manage the partners once found. We have discussed the search in terms of brokering and boundary spanning. While brokering and boundary spanning both suggest that a firm is accessing sources that are somewhat distant, this distance can often be rather limited. Thus, a key issue in relation to open innovation is how to find the sources of external knowledge that may be valuable and not restrict the search to local networks (Meulman et al., 2018). As we have discussed previously, innovativeness depends on introducing novel knowledge to create new products or services. The idea of open innovation is that this novel knowledge is more likely to reside outside the firm boundaries. However, the focal firm has to identify the potential sources of innovative but relevant knowledge.

The concept of absorptive capacity alerts us to the likelihood that an organisation will identify more local search partners (organisations that have a technical and market knowledge base similar to the focal firm) than distant search partners (organisations with knowledge that could be useful but is far removed from the focal firm's current knowledge base). This creates the problem that Meulman and colleagues describe as the local search trap. This trap is compounded by the fact that rationality is bounded (as we have previously seen), so that managers tend to rely on what and who they know, and also by the availability heuristic. The availability heuristic explains that experiences in the past and experiences that are more alien are less likely to be recalled than experiences that are recent and familiar, meaning that managers may ignore more distant partners even if they are vaguely aware of them. All of these limitations, strongly associated with the concept of absorptive capacity, mean that when a company is looking for potential open innovation partners, it will often restrict itself to familiar, proximate knowledge areas (Almeida et al., 2011).

Almeida and colleagues suggest that one solution to this local search problem is to use a search engine that can be specifically designed to include distant as well as local knowledge while making the search efficient. Based on these criteria, they developed a web-based search tool that they prototyped with different companies. This showed how such a digital tool could dramatically expand the range of open innovation partners that a focal firm considered – well beyond the personal networks of individual managers, even if they had quite extensive networks that included many structural holes. For example, an event management company was looking to create more sustainable solutions for the festivals that it put on, especially in relation to providing energy solutions at remote sites. Using the digital search tool, the manager was able to identify 47 potential partners. While 20 of these were already familiar to the company, 27 were new, including potential partners from new sub-sectors that had not previously been considered (e.g., shipbuilding companies that produce batteries for autonomous energy systems on boats but that could also be used at remote-site festivals). Moreover, the 27 included some potential partners with more local knowledge – indicating that relying on personal networking and searching is suboptimal even compared to finding more proximate knowledge partners.

Open innovation and governance

While we have just discussed that finding appropriate open innovation partners is one key issue, another is managing and governing the relationships between partners once established – whether these partners are individuals involved in a competition or companies working in an alliance. In this regard, Felin and Zenger (2014) address the question of when firms should use specific open forms of governance but also when more closed forms would be better. Specifically, these authors consider how the attributes of a problem needing to be addressed in an innovation project should influence the governance choices. They suggest that there are two key dimensions to consider in relation to the problems that need to be addressed to progress an open innovation project – the complexity of the problem and the amount of hidden knowledge that is involved. In terms of complexity, this relates to the number of different aspects that need to be solved, together with how much interdependence there is between these. Where there are a lot of different aspects that are highly interdependent, the problem is a complex one. Where there are only a few different aspects to the problem and each of these aspects can be done more or less independently, then the problem is simple. Think, for example, of an innovation project to develop a self-driving vehicle (many different aspects and lots of interdependencies) as against an innovation project focused on starting up a window-cleaning service (only a few basic problems to solve and mostly independent).

In terms of the amount of hidden knowledge involved, this relates to how far the knowledge that is needed can be identified in advance. For some problems, it is clear what knowledge is needed and this knowledge can therefore be specifically sought either internally or externally. For other problems, what knowledge might be useful is more obscure, so the focal firm needs to motivate those with potentially relevant knowledge to get involved rather than simply ask for the involvement of particular individuals or organisations.

These two dimensions then influence the type of governance model adopted, as depicted in Table 6.2.

From this analysis, then, open innovation is least likely to be an option where the complexity of the problem is high and the knowledge that is needed well known. In these cases, such as in the Defence-co InnovationJam, the innovation can best be done as a liminal space, either in-house or through alliance partners that can work closely together as will be needed given the interdependencies between the different parts of the solution. For simpler problems but again where the needed knowledge is well known in advance, contracts can be written for external individuals or organisations to undertake specific pieces of work that will then be integrated by the focal firm.

When there is considerable hidden knowledge – that is, it is unclear in advance what knowledge will be useful, then more open forms of innovation

Table 6.2 Selecting different open innovation strategies (adapted from Felin and Zenger, 2014)

Problem complexity			
Hidden knowledge		**Simple**	**Complex**
	Low	Contracts	Alliances or closed innovation
	High	Platforms and Contests	User-directed open innovation

can be used. This was explored with the city of Vienna's co-creation project, in which it opened innovation through an idea contest to attract a wide breadth of potential knowledge to benefit the new tourism strategy. For problems where the complexity is simple, using online platforms to reach a wide audience (and so identify sources of knowledge that may have been missed) and then setting up tournaments that can result in very diverse solutions is a good open solution. On the other hand, if the problem is highly complex, then such competitions may be problematic because different parts of the solution will not easily fit together. In these situations, user-led open innovation projects are the preferable form of governance since the users have strong motivation to work together to solve problems that they themselves identify.

Conclusions

In this chapter, we have discussed project liminality and open innovation, the differences between these concepts, and when they can be used in digital innovation initiatives in organisations. This is important in understanding why liminal spaces are created within organisations in order to separate innovation from typical organisational routines or, in some cases, more radical innovation is needed in the form of open innovation where knowledge is sought from external stakeholders, such as customers, clients and even competitors. In open innovation, it is important to decide what strategy to adopt and how to ensure the mutual advantage of those involved. It is also important to think about how to use digitally facilitated networks to access diverse knowledge and how to govern the process of innovation. A contingency approach is helpful here to understand how the complexity of the project and determinacy of the knowledge needed will influence these choices.

The following discussion questions are relevant to using this chapter in teaching exercises and discussions or for revision:

1. What are the key differences between liminal projects and open innovation? Give examples of digital innovation projects to illustrate.

2. Explain the different types of open innovation arrangements. In which situations of digital innovation might each arrangement be used?

3. What are the differences between digital product and service innovation? Give an example for each where open innovation may be suitable.

4. What are the key factors to consider in deciding on an open innovation approach for a digital innovation project?

 ## Case questions

The following case questions might also be relevant to using this chapter in teaching exercises and discussions:

1. What might be the reasons Defence-co chose to use IBM to help structure and moderate its innovation project?

2. While the city of Vienna's co-creation project was seen as a success, senior management decided not to continue with large-scale open innovation in the future. What reasons might there be for this?

 ## Additional suggested readings

Burt, R. S. 1992. *Structural Holes: The Social Structure of Competition*. Cambridge, Massachusetts: Harvard University Press.

Chesbrough, H. W. 2003. "The Era of Open Innovation," *MIT Sloan Management Review* (44:3), pp. 35–41.

Chesbrough, H. 2017. "The Future of Open Innovation,". *Harvard Business Review* (60: 6), pp. 29–35.

Chesbrough, H., and Appleyard, M. (2007). "Open Innovation and Strategy," *California Management Review* (50:1), pp. 57–76.

Felin, T. and Zenger, T. (2014). "Closed or open innovation? Problem solving and the governance choice," Research Policy (43), pp. 914–925.

Cohen, W. M., and Levinthal, D. A. 1990. "Absorptive Capacity: A New Perspective on Learning and Innovation," *Administrative Science Quarterly* (35:1), pp. 128–152

Fleming, L., and Waguespack, D. M. 2007. "Brokerage, Boundary Spanning, and Leadership in Open Innovation Communities," *Organization Science* (18:2), pp. 165–180.

Gupta, S., Woodside, A., Dubelaar, C., and Bradmore, D. 2009. "Diffusing Knowledge-Based Core Competencies for Leveraging Innovation Strategies: Modelling Outsourcing to Knowledge Process Organizations (Kpos) in Pharmaceutical Networks," *Industrial Marketing Management* (38:2), pp. 219–227.

Marabelli, M., and Newell, S. 2012. Knowledge risks in organizational networks: The practice perspective. *Journal of Strategic Information Systems* (21:1), pp. 18–30.

Obstfeld, D. 2005. "Social Networks, the Tertius Iungens Orientation, and Involvement in Innovation," *Administrative Science Quarterly* (50:1), pp. 100–130.

Porter, M. E. 2008. "The Five Competitive Forces That Shape Strategy," *Harvard Business Review* (86:1), pp. 25–40.

Trkman, P., and Desouza, K. C. 2012. "Knowledge Risks in Organizational Networks: An Exploratory Framework," *The Journal of Strategic Information Systems* (21:1), pp. 1–17.

Vasudeva, G., Zaheer, A., and Hernandez, E. 2013. "The Embeddedness of Networks: Institutions, Structural Holes, and Innovativeness in the Fuel Cell Industry," *Organization Science* (24:3), pp. 645–663.

References

Almeida, P., Hohberger, J., and Parada, P. 2011. Informal Knowledge and Innovation. In *Handbook of Organizational Learning & Knowledge Management*. Eds. Easterby-Smith, M., and Lyles, V.A. Chichester, UK: John Wiley & Sons.

Bhalla, G. 2010. *Collaboration, Co-Creation, and Innovation Jams: An Interview with IBM's Liam Cleaver*. Bhalla: Keeping Innovation Flowing, Available at: http://www.gauravbhalla.com/collaboration-co-creation-and-innovation-jams-an-interview-with-ibms-liam-cleaver.

Bjelland, O.M., and Wood, R.C. 2008. "An Inside View of IBM's InnovationJam," *MIT Sloan Management Review* (50:1).

Chesbrough, H. 2010. "Bringing Open Innovation to Services,". *MIT Sloan Management Review* (52: 2), pp. 85–90.

Chesbrough, H. (2017). "The Future of Open Innovation,". *Harvard Business Review* (60: 6), pp. 29–35.

Chesbrough, H., and Appleyard, M. (2007). "Open Innovation and Strategy," *California Management Review* (50:1), pp. 57–76.

Chesbrough, H. W., and Garman, A. R. 2009. "How Open Innovation Can Help You Cope in Lean Times," Harvard Business Review (87:12), pp. 68–76..

Czarniawska, B., and Mazza, C. 2003. "Consulting as a Liminal Space," *Human Relations* (56:3), pp. 267–290.

Earl, M.J. and Skyrme, D.J. 1992. "Hybrid Managers: What do we know about them?". *Journal of Information Systems*, (1:2), pp. 169–187.

Felin, T. and Zenger, T. (2014). "Closed or open innovation? Problem solving and the governance choice," *Research Policy* (43), pp. 914–925.

Garsten, C. 1999. "Betwixt and Between: Temporary Employees as Liminal Subjects in Flexible Organizations," *Organization Studies* (20:4), pp. 601–617.

Hagen, R., Miller, S., and Johnson, M. 2003. "The Disruptive Consequences' of Introducing a Critical Management Perspective onto an Mba Programme: The Lecturers' View," *Management Learning* (34:2), pp. 241–257.

Hendry, J., and Seidl, D. 2003. "The Structure and Significance of Strategic Episodes: Social Systems Theory and the Routine Practices of Strategic Change," *Journal of Management Studies* (40:1), pp. 175–196.

Howe, J. 2009. *Crowdsourcing: How the Power of the Crowd is Driving the Future of Business*. London: Random House.

Lave, J., and Wenger, E. 1991. *Situated Learning: Legitimate Peripheral Participation*. Cambridge: Cambridge University Press.

Marabelli, M., and Newell, S. 2012. Knowledge risks in organizational networks: The practice perspective. *Journal of Strategic Information Systems* (21:1), pp. 18–30.

Meulman, F., Reymen, I., Podoynitsyna, K. and Romme, G. (2018.) "Searching for Partners in Open Innovation Setting: How to overcome the constraints of local search," *California Management Review* (60: 2), pp. 71–97.

Morton, J., Wilson, A., and Cooke, L. 2016. Open Strategy Initiatives: Open, IT-Enabled Episodes of Strategic Practice. *Pacific Asia Conference on Information Systems (PACIS)*. Chiayi: Taiwan.

Osterwalder, A., and Pigneur, Y. (2010). *Business Model Generation: A Handbook for Visionaries, Game Changers, and Challengers*. Hoboken, NJ: John Wiley & Sons.

Rottenburg, R. 2000. "Sitting in a Bar 1," *Studies in Cultures, Organizations and Societies* (6:1), pp. 87–100.

Royer, I. (2003). "Why bad projects are so hard to kill,". *Harvard Business Review*, February.

Schlagwein, D., and Bjorn-Andersen, N. 2014. "Organizational Learning with Crowdsourcing: The Revelatory Case of Lego," *Journal of the Association for Information Systems* (15:11), pp. 754–778.

Tempest, S., and Starkey, K. 2004. "The Effects of Liminality on Individual and Organizational Learning," *Organization Studies* (25:4), pp. 507–527.

Turner, V. 1969. *Liminality and Communitas*. Chicago, IL: Aldine.

Turner, V. 1982. *From Ritual to Theatre: The Human Seriousness of Play*. New York: PAJ Publications.

Van Gennep, A. 1909. *Les Rites de Passage*. Chicago, Il: Chicago University Press.

Wagner, E. L., Newell, S., and Kay, W. 2012. "Enterprise Systems Projects: The Role of Liminal Space in Enterprise Systems Implementation," *Journal of Information Technology* (27:4), pp. 259–269.

Wenger, E. 1998. *Communities of Practice: Learning, Meaning, and Identity*. Cambridge: Cambridge University Press.

7 THE ROLE OF OBJECTS IN ORGANISING FOR DIGITAL INNOVATION

Summary

In this chapter, we discuss the role of material objects in processes involving knowledge creation/sharing and in facilitating innovation. We open the chapter with a case on healthcare innovation. The case, which examines a project aimed at improving coordination of care, and the remainder of the chapter are illustrative of practices involving a number of objects (PowerPoint presentations, medical sheets and other documents) that allow knowledge-sharing between the various people involved, often belonging to different occupational communities. Throughout the chapter, we explain the role of objects in organising and we discuss the idea that objects (as well as people) have agency (i.e., make things happen) in relation to innovation processes. In particular, we discuss strategic objects (used by managers to share ideas with their peers) and boundary objects, a more emergent and bottom-up view of how objects, through use, become relevant actors during innovation processes. We also consider the role of emotions in humans engaging with objects.

Learning Objectives

The learning objectives for this chapter are to:

1. Understand the role of objects as actors that can 'make things happen' in their own right and not just as tools of human actors

2. Consider the differences between material and human actors during the innovation process

3. Understand the difference between strategic objects and boundary objects

4. Consider the role of power during innovation processes

5. Recognise the emergent nature of boundary objects and the idea that these objects have interpretive flexibility but can also have material 'plasticity' that allows them to endure over time

6. Understand the idea of social/material 'imbrication' as a way of conceptualising how material objects change practice over time but also are themselves changed as practices change.

Canada-Care case: The role of objects in an innovative healthcare initiative in Canada

The case concerns a pilot project that was subsequently transformed into a hospital and province-funded programme, aimed at improving coordination of childcare interventions at Canada-Care (a healthcare network in Canada), specifically within the hospital (where the pilot/project was housed) and between other healthcare and social service providers. The initiative focuses on children with complex care needs, that is, children with multiple and life-threatening symptoms requiring treatment by several specialists. The need to improve healthcare coordination emerged in 2008, when a number of the children's parents expressed concern that different healthcare specialists (within the hospital) were not exchanging crucial medical knowledge. As a result, the parents were often overwhelmed and emotionally drained because it fell to them to coordinate the care of their child, even though their lack of knowledge of medical terms might lead to imprecision in reporting their child's circumstances. The families were particularly vocal about their issues during 'family forums', monthly meetings held at the hospital and involving parents and carers. An additional coordination issue was that the 'external' providers were not always aware of each child's most recent health issues, and this too posed health risks. Further, the hospital did not receive the most recent updates from school or social service agencies about the children's social/psychological condition (Figure 7.1).

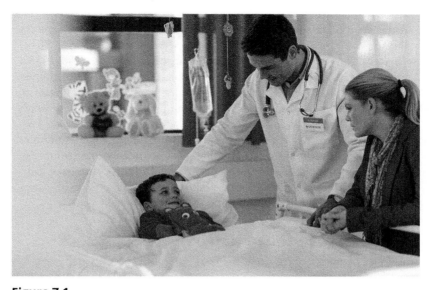

Figure 7.1

At Canada-Care, the clinician's engagement was paramount for improving healthcare service delivery
©Getty Images/Westend61

The project analysis and bidding/approval process took two years (2008–2010) and included a major literature review on coordination and the identification of various initiatives (in the USA and the UK) aimed at improving paediatric coordination. The project in its pilot form was initiated by a core team that included six individuals who had various clinical and administrative roles at the hospital and with external providers. The Canada-Care pilot received grant funding in 2010 for two years and included three external providers (social services, access to care and psychological support) who were also involved in the bidding process. Additional funds were secured in 2012 for a further two years. In 2014, permanent funding was secured (from the hospital and the province) and the project became an established programme.

Initially, 20 children were enrolled. A nurse coordinator was hired to be responsible for facilitating communication between clinicians and families and one of the hospital doctors (a member of the core team) undertook the role of full-time coordinating physician. The doctor's main task was to engage various hospital specialists, as well as external/independent paediatricians, in changes associated with communication, knowledge-sharing and the like. A hospital's registered nurse, who was also a member of the core team, took the role of project manager. A healthcare manager (another member of the core team) became the pilot project's director. Steering and advisory committees were created. In spring 2014, the transition from pilot to programme involved hiring an additional nurse coordinator (full-time) and three physicians (one full-time and two part-time), which increased capacity to enrol 40 more children. Moreover, four additional providers joined the programme: palliative care, a youth community centre, a respite centre and an additional social services provider. For the various partners, being part of the pilot (or programme) involved frequent interactions including steering and advisory committee meetings but also informal meetings and small group conference calls. Because of the involvement of a number of doctors, together with carers from outside the hospital, coordination was a major challenge.

Objects used in the initiative

A number of objects were used, with various purposes, over the years. For instance, PowerPoint and poster presentations were used by physicians to communicate with nurses and social workers, as well as with specialists within the hospital, to highlight benefits from coordination and to suggest new practices to achieve it. Presentations were also used in steering and advisory meetings, where the providers involved illustrated existing coordination issues (related to their specific organisation). PowerPoint presentations helped sharing across occupational communities by embedding knowledge that could be interpreted in various ways.

Research documents on coordination (reviews and various studies and executive reports) represent another example of objects that were helpful to support knowledge-sharing. For instance, a master's student undertook a dissertation on healthcare coordination during the period 2007–2008. The student's literature review in their dissertation, along with comparisons of research on successful coordination projects, was examined by the doctors, nurses and social workers. All these people were able to understand the relevant message inscribed in this document that was related to past successful practices used to improve coordination of care in networks.

One of the most effective objects, however, was a medical sheet which, like those that used to be found on a clipboard at the foot of a patient's bed, was shared across the various agencies and among specialists within the hospital. The medical sheet was updated regularly by a hospital doctor. The updates included deleting information no longer relevant and adding new information. The sheet was initially suggested by a parent who had developed his own 'cheat sheet' to help him remember his child's current details when visiting specialists. Once updated, the medical sheet would be provided to the parents, so they had it with them when visiting different specialists, and, if needed, to the Emergency Room (ER) and was also sent via fax or email to relevant departments within the hospital and to external organisations. In the original project proposal, the goal had been to develop an

Electronic Medical Record System (i.e., a digital innovation) to allow sharing across the various people involved in the care of a child. However, the resources necessary for this and the time it would take for this digital system to be developed meant that this plan was abandoned in favour of using the simple paper-based system to share information.

While the purpose of the medical sheet was to share information among those involved in the child's care, in practice it also provided relief to parents who no longer needed to understand complicated medical terms or very long clinical reports. It included only relevant (e.g., about a life-threatening condition) information about the child in a way specific enough to be meaningful for doctors but easily understandable by non-clinicians (e.g., schools, social workers, and families). Over time, the sheet was modified (e.g., social services suggested adding fields related to behavioural issues, while the medical sheet started out as a 'clinical' sheet, then some non-necessary clinical details were replaced by more social and psychological data).

The introduction of the medical sheet was associated with changes to other practices. For instance, its introduction in 2010 led to fewer visits to the ER since the nurse coordinator, relying on the updated information, could answer most of the families' questions without their having to rush to hospital. Moreover, when parents did take their child to the ER, they found that with the medical sheet in-hand, they were listened to by the clinicians because the sheet, which was 'signed by a doctor and not by a mum', gave them a legitimacy that they had not had when previously they had tried to quickly explain the treatment that they knew their sick child needed. Another example of change concerns the possibility for families to travel with their children, dramatically changing the quality of family life. For instance, in 2012, one family was able to go abroad (to the USA) on vacation with their child. Along with other arrangements, the medical sheet was sent to three hospitals in Orlando, where the family was planning to stay for a week. A further example concerns the number of meetings between the coordinating physician and external agencies, which in 2013 decreased from as many as twice a month to just three or four times a year. More importantly, discussions focused more on how to further improve coordination rather than on specific cases since these were mostly being dealt with via the medical sheet's ongoing use. Figure 7.2 depicts the medical sheet that was actually used at Canada-Care (we removed some details of the sheet to keep the case disguised).

All the objects described here (e.g. PowerPoint slides and various documents, including the literature review and the medical sheet) were able to help with knowledge-sharing across occupational communities, thereby leading to the creation of innovative practices. These practices involved a better understanding of the relevance of coordination to improve care (specialists, through PowerPoint slides delivered by the coordinating physicians); awareness of coordination techniques (from past initiatives) that promoted the creation of similar practices in the network examined in this case; and distributed knowledge on patients across various carers but also involving

Single Point of Care "SPOC" Document

Coordination of Care Pilot Project Participant's Information Summary

Allergies/Reaction: _____

Caregiver's Names: _____

Legal Guardian: _____

   Married Separated Divorced Foster care Single parent

Emergency Contact Information _____

Code Status: _____

Language Spoken _____

Patient Identification

Family Physician/Pediatrician: _____ School: _____

Community Pharmacy: _____ Telephone: _____ Fax: _____

Family Coordinator: _____ Telephone: _____ Fax: _____

Diagnoses:

Brief Medical History:

_____
_____
_____
_____
_____
_____
_____

Current Medication List:

Medication/Natural remedy	Dosage	Reason for the Medication

Routine Immunizations Up to date: yes   no (for more details see "Hospital Immunization Record" Form No. 1049)
Special Vaccine details:
_____

List of Physicians and Programs following at CHEO:

Name:	Specialty:	Last Visit:	Next Visit:

Figure 7.2

The medical sheet used at Canada-Care

the families, who could access (and carry) the medical sheet. The sheet was constantly updated by clinical and administrative staff – often upon the families' suggestions. Indeed, two representatives of the families sat on the advisory committee. The medical sheet template has changed over the years, not just to accommodate carers' needs but also to meet requests coming from the families who, having been actively involved in the creation and implementation of this document from the outset, now had a say in what should be highlighted therein.

We had a family member whose husband was incarcerated. And so mum actually wanted that in the document under relevant family history. She felt that it was important that people knew that the father wasn't involved because people often will ask.

Summary

This case illustrates how several objects were able to 'actively' play a role in the innovation processes undertaken by the healthcare project in Canada-Care. Most of these innovation processes involved knowledge-sharing (the creation of new knowledge that built on existing experience of the people involved). However, some objects were purposely created to promote innovative practices (e.g., the PowerPoint presentations) while others emerged as helpful devices to promote discussion around how to change healthcare practices (e.g., the literature review and the medical sheet). The medical sheet was also modified over the years – promoting novel ways to share knowledge yet maintaining intact its main functionalities (a centralised document containing key health information about a child with complex care needs). Eventually, the paper-based system was also converted into a digital medical record-sharing system. The case is, as with previous chapters, used to illustrate and discuss key topics and themes.

Introduction: Agency, objects and organising

This chapter explores the importance of considering the role of objects in understanding how innovation processes can unfold over time. Here, we discuss how the recent literature has acknowledged the importance of focusing not just on how people perform actions, in practice, but also on the role of material 'things', which themselves perform actions. In particular, here, we emphasise that objects support knowledge-sharing and knowledge creation processes and ultimately innovation. Think of enterprise systems, which are used by several individuals variously for purposes related to sharing organisational data, creating reports and prospects but also translating knowledge into forms that are more easily understood by people with different backgrounds and business needs. For instance, an enterprise system generates transactional data (everyday processes involving, for example, creating invoices, performing payments and issuing goods receipts) that originate through practices involving employees (in the accounting function, say), customer and supplier organisations (which send/receive invoices and payments and which ship and purchase goods), computers (hardware), networks and software. These data are then elaborated upon and manipulated by systems, software/algorithms and people and a 'new' version (or better, a selection) of the same data is presented in the form of synthesised reports to managers and executives, who use these data to make (strategic) decisions.

This example illustrates that knowledge creation and sharing processes occur through joint practices involving people and objects. Furthermore, the knowledge originating from transactional data (thousands of everyday transactions) is translated in ways (e.g., reports) that make it more easily understood by the recipients.

Other examples might involve different and more concrete objects such as written documents, a medical sheet and a PowerPoint presentation, as in the Canada-Care case. In addition, the knowledge that emerges through the use of these objects can be shared with people within the same 'community'. For instance, drawing again on the case, a medical sheet can be helpful in hospital settings where several physicians are caring for a particular patient. Or it can be translated across different occupational communities. The same medical sheet that informs doctors can remind the parents of patients about what medications their child needs to take and who (which doctor) is currently overseeing their treatment. This example of medical sheets can be extended to digitally based systems such as electronic medical record (EMR) and electronic health record (EHR) systems. These systems allow storing, exchanging and updating clinical information within a hospital and across a network (e.g., as in the Canada-Care case, between the various agencies involved in the pilot project) (Figure 7.3).

Figure 7.3

Presentations, for instance, included objects such as PowerPoint, which was proven helpful to translate knowledge across practitioners
©Getty Images

Objects and innovation processes

The idea underpinning the relevance of objects in these knowledge-sharing and innovation processes is that, as well as people, objects have agency. They *do* things. In other words, objects, as well as people, have agency – called material and human agency respectively (Leonardi, 2011). The idea that objects have agency is relatively new and stems from the progressive acknowledgement, in organisational and management information systems studies, that technologies (especially digital devices but also other technologies that convey information, including paper) are not just tools that can be exploited by organisational actors. The beginnings of this approach can be traced back to the 1970s, when sociotechnical scholars began to elaborate on the idea that organisational processes should be seen from a perspective that involves people *and* technology jointly (not just people who command technology). This realisation emerged from a series of studies that showed how the ways people and technology were organised together influenced outcomes. For example, when new technology was introduced to coal mining (machinery being introduced to cut the coal and move it on a conveyor belt to the surface), the decision was made to organise workers into shifts, each shift responsible for a part of the mining cycle, from clearing the face ready for cutting, through to getting the coal to the surface. This 'longwall' method was very different from the traditional 'shortwall' method, where small groups of workers worked cooperatively to complete the entire mining cycle, using mostly hand tools. The shift work introduced with the new technology disrupted the social system where the small teams had supported each other in the dangerous conditions of a mine and where doing the complete cycle as a team meant that they could flexibly adapt to the different conditions of the coal surface. The new, longwall method resulted in a lot of unrest among the workers and manifested in absenteeism and blaming of each other when one shift came to work but could not proceed because the previous shift had not fully completed its task. The research team from the Tavistock Institute (a research institute involved in social science research based in the UK) concluded that a sociotechnical perspective was needed. This proposed that the social system needed to be designed to fit the technical system in order to maximise productivity – which had to include the social and psychological effects of the selected organisation. In this instance, they introduced the 'composite longwall' method, where each team had responsibility for composite tasks so that the shifts were not each working on a specialist part of the process and could therefore be able to adapt to the changing conditions of the mine. The responsibility given to the shift team also meant that they became more supportive of each other, important in this dangerous type of work.

Objects, agency and organising

Scholars made a further step forward (in the 1990s and early 2000s) when they acknowledged the 'agentic' role of material objects in everyday practices (Pickering, 1993; Suchman, 2007). Generally speaking, the practice perspective goes beyond sociotechnical approaches because the focus shifts from the mere identification of relationships *between* humans and things to the practices that *emerge* from these relationships. In other words, from a practice perspective, human and material agency do not simply interact but are instead interlocked – our 'doings' are always represented by practices involving material as well as social aspects. Therefore, people do not exploit or simply use objects but instead *enact* them in practice. The word enactment is used by practice scholars to highlight that the relationship between people and objects represents a mutual engagement rather than a one-off exploitation or a mere use of a technology/object.

Orlikowski (2007) offers a very practical example of the entanglement of humans and objects, both having agency. The example concerns Google (the search engine) and illustrates how the everyday use of digital artefacts reveals their agency and how they relate with each other. People use Google on computers (or they *google*) to search pages and find information and knowledge. The search engine returns results in a certain order on the basis of an algorithm (PageRank, patented by the Google founders when they were in college), which dynamically modifies the ranking of the various results, from the most relevant to the least relevant. The Google algorithm does so on the basis of previous searches (by humans). Moreover, *googling* from different locations or with a smartphone or a tablet leads to different results because the material agency of the Google algorithm adapts to humans and, in turn, takes actions that might affect what humans think and do. The underlying algorithms also provide customised ads on the basis of how people use these platforms. This in turn has the power to modify people's *doings*; a social media user might make purchasing choices (more or less informed, this is debatable) in response to the personalised ads they see (see Chapter 8 for more discussion on social media).

Summary: Objects in knowledge-sharing and digital innovation

In this chapter, we focus on two main ways in which objects support knowledge-sharing and innovation. First, we examine so-called strategic objects. In organisational settings, these objects are purposely created by 'strategists' (also see the discussion on strategy-as-practice in Chapter 4) to share strategic considerations among senior management executives and other stakeholders. These objects are relatively static but can be extremely effective to promote the sharing of ideas. Second, we examine objects that are either set up or emerge during everyday organisational life – known as *boundary objects*. Boundary objects are particularly relevant to knowledge-sharing and innovation because they incorporate knowledge in a form that is understandable by different occupational communities.

Both types of objects (strategic objects and boundary objects) are present in the Canada-Care case that opens this chapter. Strategic objects were created by doctors (e.g., the PowerPoint presentations) and are subsequently enacted in practice during presentations and meetings; boundary objects emerged (e.g., the literature review) and were subsequently used by other actors. The medical sheet is another boundary object whose boundary-spanning effectiveness emerged through use by the carers to share medical knowledge across the healthcare network.

In sum, understanding the role of materiality (objects) and how human and material agency interact in organisational and networked settings is relevant to make sense of how knowledge and innovation are collective processes where 'things' should be considered just as relevant as people. We next discuss strategic objects in more detail, then move to boundary objects and conclude by providing insights on how emotional engagement in the use of objects can amplify the collaborative effects instigated by various objects (both strategic and boundary). While from a theoretical perspective people and objects are equally important (i.e., they have agency), in reality human agency has emotional components, which is not the case for objects themselves.

 Key Concepts: Sources of Agency

Human Agency:
Organisational actors who, more or less intentionally, participate in the making of organisational life by doing things such as making phone calls, participating in meetings, interacting with other human agents and with 'things'

Material Agency:
Any actors that are not human actors. These include physical objects (e.g., a desk, chair or computer) and the meanings inscribed in them. Material actors (since they have agency) 'do' things. For instance, a more or less comfortable office chair affects our posture when we work; the sound of a fire alarm makes us head for the emergency exit

Strategic objects

As explained in Chapter 4, the strategy-as-practice approach emphasises the day-to-day activities of practitioners who shape, refine and actualise strategy through what they do. A strategy is viewed as an emergent set of practices, which are constantly in-the-making (Jarzabkowski, 2005). Most of the recent strategy-as-practice literature conceptualises materiality as a set of strategic objects or tools (Spee and Jarzabkowski, 2009). Such objects generally refer to very common office entities (a desk, a computer, 'post-its', block notes etc.) as well as more sophisticated digital entities such as PowerPoint presentations (Kaplan, 2011) or other visual tools (e.g., pictures, videos) that are used to deliver ideas but that can also incorporate a symbolic value (Paroutis et al., 2015).

During steering committee meetings, workshops and the like, strategists (generally senior executives) are often observed using a variety of objects to facilitate others' understanding of their (strategic) ideas, to promote collaboration and knowledge-sharing or to support their own (strategic) claims in an attempt to execute their own agenda. These strategic objects are often used constructively. For instance, Gantt charts and PERT diagrams (see Chapter 5) can be used to present the desired outcome of a process and give a shared view of the progress of an innovation project. Furthermore, the timeline of a Gantt chart has a symbolic meaning related to the urgency to accomplish a project (Spee and Jarzabkowski, 2009). The strategic component embedded in these objects relates to their ability to engage team members in meeting deadlines, being productive while collaborating and the like. In addition, these objects are generally created when a new innovation project starts, as obviously a Gantt chart needs to be project-specific. To this end, as we will see below, these and other types of strategic objects have a sort of 'expiration date' because they can be used as long as the purpose they serve is still something to pursue. Once a project is over, the related Gantt is useless.

Other examples of these types of object that expire (after an objective is achieved or is abandoned) involve so-called strategic maps, that is, roadmaps that executives craft and discuss with top management to address changes, clarify high-level objectives and highlight the link between a certain strategy and the creation of value for customers. For instance, in a recent *Harvard Business Review* article, Gary Pisano (2015, p. 50) explains that 'Unless innovation induces potential customers to pay more, saves them money, or provides some larger societal benefit like improved health or cleaner water, it is not creating value'. Therefore, in order to execute a top-down strategy where top management creates an idea for an innovative product or service and then this idea is further developed (by the R&D department or through open innovation), promoted (by marketing) and successfully commercialised (by sales), it is paramount to communicate the idea to all stakeholders involved. For example, a roadmap can be used to help communicate to the

Figure 7.4

Strategic tools such as Gantt and PERT charts are used in most business meetings
©*Getty Images/Cultura RF*

various department heads involved who will then be able to supervise the various stages of the development of the new product or service, through to the release to the market and beyond. These strategic objects, which are often digital, are helpful to convince or persuade people that an idea needs to be taken further and to encourage the coordination of activities. Overall, they are extremely useful objects because they help sharing knowledge within a community. For instance, if we refer to the Canada-Care case, the PowerPoint slides were helpful for the coordinating physicians to share knowledge with the various specialists (Figure 7.4).

Alternative views on strategic objects

Other ways to use strategic objects can be more subtle. For instance, Courpasson (2000) presents a study on the 'Power of managerial indicators and impersonal systems of governance' conducted in the 1990s. The 'indicators' mentioned in the study are established (and legitimised) ways in which management understands organisational performance, revenues and success (e.g., a growing ROI, which indicates that, in the short term, a firm is making profits based on previous investments). However, in the specific case discussed by Courpasson, management makes decisions that are not in line with the interests of its stakeholders, who are shown the graphs. To this end, a graph on the growing short-term performance can be given meanings that overrate the wealth of an organisation, therefore legitimising (with stakeholders) strategic decisions (e.g., an acquisition) that may not be in line with long-term objectives but instead are pursued because someone at the executive level is advancing her/his own strategic agenda (perhaps because she/he aims to leave the company shortly and her/his golden parachute is related to the consolidated revenues of the firm at the time of leaving).

In contrast with the previous use of strategic objects to share knowledge among executives (cf. Pisano's *HBR* article), this subtle way of exploiting dominant positions, such as among those who sit on an executive board, through the use of objects (in the example above, a digitally produced graph that represents only part of the 'story' about how a company is doing) clearly may not foster long-term sustainable innovation. This also recalls what Cynthia Hardy calls 'process power', the power that upper management exercises when it attempts to obtain approval by others in order to advance personal interests. These interests can conflict with the interests of the organisation. Indeed, the main definition of process power (Hardy, 1996, pp. S7–S8) reflects its ability to allow 'the more powerful actors to determine outcomes from behind the scenes through the use of procedures and political routines ... and protect the status quo by mobilising the biases that are embedded in existing decision-making processes'. However, Hardy also notes that process power can be used to bring about change 'by extending access to decision-making arenas and agendas'. Overall, process power manipulates decision-making processes through the inclusion/exclusion of various actors (human, such as people in committees, and material such as documents, pieces of information and the like). Therefore, it is important to note the relationship between process power and the use of objects, in particular from a strategic perspective, because it illustrates that objects can be used in ways that can stimulate innovation processes, but also prevent innovations from being developed, should they compromise, for example, the interests of those in dominant positions at the upper management level. So, a manager may decide not to include in a PowerPoint presentation to investors certain information that detracts from the particular strategic plan that this manager favours. This shows how objects play an active role in shaping organisational dynamics in that they have agency and therefore they just *do*, regardless of whether they promote or prevent innovation.

The effectiveness of strategic objects

Clearly, the Canada-Care case presented in this chapter makes use of strategic objects in more constructive ways (in contrast to the self-interests just discussed). Indeed, the PowerPoint presentations were ad hoc crafted for specific audiences (the specialists) and aimed at delivering a clear message about new practices to be implemented in order to improve coordination across carers. These objects have been proven to be extremely effective because they 'make things happen' (Leonardi et al., 2012). Jarzabkowski and Kaplan (2015) describe the innovative power of 'strategic tools in use'. They provide a vignette about a company (Com) where top management decided to adopt a bi-dimensional tool called 'aggregate project plan' (Wheelwright and Clark, 1992). The tool was meant to help make strategic decisions on various ongoing organisation-wide projects involving several managers from different departments. One requirement to use the tool was, however, that all managers involved had the same understanding of the various projects in terms of their (high or low) impact on revenues, short- or long-term ROI and so on. The tool failed to create consensus among top management at Com on how to enact strategically the various projects because the people involved in decision-making processes viewed the projects from different perspectives. So, top management introduced different (more dynamic) tools enabling each manager to express her/his opinion on each project first. Only once consensus was reached on the strategic value of each project were managers required to make (joint) decisions on how to modify these projects to make them more effective for the whole company. The vignette shows the need to modify objects in use in order to increase their agency in making things happen, just as it would also be important to modify social arrangements if they were found to be not supporting an innovation project (e.g., if a project manager failed to win trust of his/her team).

To summarise, one can identify a set of 'typical' strategic objects or tools, often digitally produced and shared, that might work in several companies (and settings). Yet, it is almost impossible to identify universal tools that fit every situation in every company. The unsuccessful adoption of a strategic tool might nevertheless be helpful to reveal issues such as a management's misalignment in understanding organisational processes (as happened in Com). As we illustrated, this failure created the opportunity to undertake decision-making processes with innovative and more fitting tools.

In the next section, we describe boundary objects that do not 'simply' promote knowledge-sharing (and innovation) only within the same community. These objects are particularly helpful to exchange ideas and promote discussion at the network level, where different occupational communities and potentially different organisations with diverse backgrounds are involved.

Boundary objects

Initially conceptualised by Star and Griesemer (1989), boundary objects are defined as follows:

> Boundary objects are objects which are both plastic enough to adapt to local needs and constraints of the several parties employing them, yet robust enough to maintain a common identity across sites. They are weakly structured in common use, and

become strongly structured in individual-site use. They may be abstract or concrete. They have different meanings in different social worlds but their structure is common enough to more than one world to make them recognisable, a means of translation. The creation and management of boundary objects is key in developing and maintaining coherence across intersecting social worlds. (Star and Griesemer, 1989, p. 393)

From this definition, we learn that boundary objects are physical objects that can be adopted by a variety of individuals belonging to different occupational communities (or worlds). These individuals would normally encounter barriers in understanding objects used by other professionals. For instance, a lawyer would have a hard time in understanding a lab report with raw results of 'blood-work'. Boundary objects are not like this. They are 'weakly structured' because they do not look incomprehensible to people who do not have specific knowledge of a particular discipline or subject. Their structure (e.g., how the knowledge inscribed in a particular object is arranged) becomes comprehensible in different ways for different audiences (they are not structured for a specific audience of blood scientists). For instance, a blood-work report that, along with 'incomprehensible' numbers, provides a table to interpret these numbers (e.g., the level of cholesterol needs to be lower than 200 mg/Dl for healthy individuals) represents a boundary object. The same blood-work report becomes 'strongly structured' when specialists look at the values not in terms of the acceptable range of 0–200 mg/Dl but in relation to other knowledge that they have, such as root causes of possible high cholesterol. This means that for them the 'acceptable range' general statement becomes more nuanced so that even within the tolerable range (<200) they are able to claim that a diabetic individual should have less than 150 mg/Dl cholesterol, as such individuals are more prone to heart diseases. In other words, the same object can accommodate various communities' needs. Therefore, they are weakly structured in common use (a doctor or a nurse can have a quick look at the blood-work report and conclude that the patient does not have blood fat issues, while a patient who receives the report via email can conclude that she is fine). But, at the same time, the very same blood-work report can be analysed in depth by people with specific medical knowledge (and so is strongly structured in individual-site use).

Types of boundary objects

Star and Griesemer describe a number of objects such as field notes, maps and specimens, all of which were helpful to those involved in the development of the Berkeley Museum of Vertebrate Zoology in California. These objects supported collaborative practices (and eventually knowledge-sharing and innovation, as we will see below with an example from Paul Carlile)

Figure 7.5

At the Berkeley Museum of Vertebrate Zoology in California, various specimens promoted knowledge-sharing between scientists
©*Getty Images/Blend Images*

related to the organisation of interdisciplinary processes at the museum, which would otherwise have been 'siloed' within occupational boundaries – boundaries that originate from the specific (and specialised) knowledge of zoologists, biologists and natural scientists – but also within the museum's management and other staff (Figure 7.5). In much the same way, a map can be a boundary object for a group with different backgrounds as they attempt to navigate in a world they are more or less familiar with. A map could be used with different interests in mind – to find insects, to walk to a peak and to identify bird habitats, for instance.

One of the most notable applications of this concept is found in the work of Paul Carlile, where he recounts in-depth research on a product development company where each organisational unit has its own specific knowledge. As a result, the units are very good at solving 'within-unit' problems but, not having an overarching view of the organisation, fall short when an inter-unit problem emerges (e.g., when two units need to collaborate in design, production and sales processes). In other words, unit-based knowledge specialisation creates problems when different 'types' of knowledge need to be shared across boundaries. Here, it might be useful to reflect again on the Canada-Care network case in relation to knowledge-sharing among different carer agencies and parents before the introduction and use of the medical sheet.

Carlile (2002) identifies three different boundaries. 1) *Syntactic* – those that arise from different technical languages being used by different functional specialists. 2) *Semantic* – which reflect the tacit aspects of knowledge shared within a specific unit and which therefore are not fully understandable. As he notes, 'even if a common syntax or language is present, interpretations are often different which make communication and collaboration difficult' (Carlile, 2002, p. 444). 3) *Pragmatic* – the boundaries that refer to knowledge that arises from practice (doing) and therefore is context-specific (for instance, the knowledge that is developed within a community of practice). This reflects the fact that people are invested in their particular practice and, as a result, people do things according to what they already know, meaning that any innovation that disrupts this existing knowledge and practice may be resisted. In this case, to cross a boundary, knowledge needs to be *translated* and not simply shared – or in Carlile's terms, 'transformed'.

The theory of boundary objects can be applied to a wide range of organisational and network contexts, where challenges associated with knowledge-sharing are very often present. If we refer again to the Canada-Care case, all the objects discussed (PowerPoint slides and documents such as the literature review and the medical sheet) can be considered boundary objects 'at large', as their materiality 'does things' by promoting knowledge-sharing and supporting innovative changes. However, the PowerPoint slides' use is limited to a certain community (physicians) where there are no real knowledge barriers. The differences between the various specialists are more associated with their initial inability (and lack of understanding of the relevance) to share medical information about their patients. So, in this particular case, PowerPoint did not have a role associated with communicating with different groups of people who, otherwise, would not be able to understand a common message. Objects such as the literature review and the medical sheet reflect how objects get used (over time) in practice. These objects include a larger array of communities (the literature review involved all carers and administrative staff, whereas the medical sheet involved the families as well as the carers). Although an object does not become a boundary object in relation to the number of communities it affects, its effectiveness can be defined as such (Yakura, 2002). All these objects have 'plasticity' (they can be interpreted differently by various occupational communities) and 'robustness' (besides the various meanings given to the document, its main properties are stable).

Comparing strategic and boundary objects

To summarise, the main difference between strategic and boundary objects relates to the extent knowledge is shared within a specific occupational community or between different occupational communities. At Com (described above), the managerial strategic tools were used to share ideas on various organisational projects. But top management, albeit coming from different departments, had similar background knowledge on how to strategically evaluate projects, and conflicts emerged on whether some projects should be considered relevant in the short or long term. Boundary objects instead focus on cross-community knowledge-sharing. For instance, a sales strategy that needs to be shared between executives, salespeople, retailers and other stakeholders (for example, supply chain and logistics) would involve boundary objects. Since innovation projects typically involve both within- and across-occupational/community knowledge-sharing they will typically involve both types of objects in different contexts.

 Key Concepts: Different Types of Objects

Strategic Objects:
Material actors that are purposely created to share knowledge and ideas within a community. They can be used in a constructive way (e.g., to promote innovative changes) as well as in a more subtle and sometimes negative way (e.g., to 'convince' people and advance a personal agenda)

Boundary Objects:
Objects that are either specifically created or emerge in practices that involve cross-occupational and community knowledge-sharing. Such objects have plasticity to adapt to local needs and constraints of the several parties employing them yet are robust enough to maintain a common identity across sites. They are weakly structured in common use and become strongly structured in individual-site use. They may be abstract or concrete. They have different meanings in different social worlds but their structure is common enough to more than one world to make them recognisable and so provide a means of knowledge translation

The innovative power of objects

As we have seen, strategy-as-practice scholars have discussed strategic objects as intentionally managed by executives to bring different stakeholders together (e.g., related to a specific change at a certain moment in time). These objects do have agency, but their life is limited (they are ad hoc created). The role of these objects relates to their ability to persuade people, involve collaborators and share knowledge within a certain community. Strategic objects are therefore key to a successful innovation process. Yet, they remain very ephemeral organisational artefacts (they are one-off and are created and executed by change agents, management and executives). This is different from some boundary objects, which can be used for a long time in various contexts. For instance, this textbook contains knowledge that (we hope) is understandable by students, instructors and professionals (practitioners). We (instructors) provide practical examples to make theorising around concepts accessible to those outside academia and therefore with a different (less theoretical but more practical) background. This textbook provides relevant knowledge on key organisational theory concepts. Yet, we attempt to present these concepts in a way that is general enough so that many different types of people can access them.

This textbook is, however, somewhat static. It can be used for several years but unless a new edition is released, the knowledge inscribed in it remains the same. Digital objects offer the opportunity to build objects that are constantly modifiable. Wikipedia is an example of a boundary object with knowledge that should be accessible to most and which can be revised constantly as knowledge on a particular subject changes. For instance, the biography of famous people alters as soon as they start a new job, divorce or die, and the relevant change can be made to Wikipedia. These updates happen almost in real time.

So, Wikipedia is a powerful actor, playing a major role in mobilising knowledge across a network. Here, a material digital object is an actor in its own right since the object is not controlled by those leading the project (the Wikipedia administrators) because once it is released into the scene it then becomes interlocked (Leonardi, 2011) in the ongoing social dynamics (everyone can access and change a Wikipedia page). The properties of the material actor are modified by those who adopt it. At the same time, though, the adopters change their practices to meet the new properties of the material actor. One way of capturing how objects and people are constantly interlocked in ongoing practices is using the metaphor of *imbrication*.

The verb imbricate is derived from names of roof tiles used in ancient Roman and Greek architecture. The tegula and imbrex were interlocking tiles used to waterproof a roof. The tegula was a plain flat tile laid on the roof and the imbrex was a semi-cylindrical tile laid over the joints between the tegulae. A roof could not be composed solely of tegulae nor of imbrices. Human and material agencies, though both capabilities for action, differ phenomenologically with respect to intention. Therefore, like the tegula and the imbrex, they have distinct contours yet they form an integrated structure through their imbrication (Leonardi, 2011, pp. 150–151) (Figure 7.6).

The concept of imbrication is helpful to understand how the interlocking relationships between human and material agencies unfold over time. Namely, if we look beyond sociotechnical systems and take a more practice-based view of the inter-relationships between human and material agencies, the metaphor of imbrication is helpful to understand how these two agencies cannot be separated since each affects the other in an ongoing 'dance' of agency. This is also helpful to explain how practices change as a consequence of the use of objects and how some objects (especially digital) can adapt to emerging practices by being modified.

Figure 7.6

Imbrexes and tegulae are interlocked, yet they are two distinct entities
©*Getty Images/iStockphoto*

Objects and everyday organising

Table 7.1 synthesises the powerful role of objects in everyday organising practices that support the development of new products and services.

We need to note that not all boundary objects are effective for a long period of time. However, the characteristics of these objects, as per Star and Griesemer's definition (they can change as long as they are 'robust' enough to maintain their key meanings), make them potentially important actors during innovation processes since they can last for a long time. This is the case because when practices change, boundary objects can be shaped by those involved to adapt to new contexts. In addition, because boundary objects can be defined as such if enacted by different communities (i.e., people with different knowledge backgrounds), they are emergent, and are boundary objects because of how they are used, in practice, by different communities. In contrast, as we previously pointed out, strategic objects are often used in the short term because they are ad hoc created (by strategists) for a specific purpose.

In a recent study (Marabelli et al., 2017), we noted that boundary objects are plastic not just in terms of their cognitive interpretations (i.e., weakly structured objects can be interpreted because they provide knowledge that is understandable by a wide range of individuals, or 'worlds') but also in terms of their physical/material characteristics. Objects that are 'materially' plastic allow actors to change them over time. For instance, the medical

Table 7.1 The power and role of objects

Object/ innovative practices	Strategic objects	Boundary objects
Agency	They do things to the extent that they support the sharing of new ideas among peers.	They do things because they change people's practices (their function is not limited to sharing ideas).
Innovative power	Their power resides in the symbolic meanings that they carry.	Their power is related to the 'change agent' role that they have across communities.
Examples of their effectiveness	A graph showing growing ROI is visually impactful because ROI growth underpins a series of consequences for the company.	They are effective to the extent that they make knowledge available to different communities.
Longevity	These objects are constantly recreated and used (PowerPoint slides, Gantt charts).	These objects are constantly reshaped by those involved, so a boundary object, potentially, can last longer than a strategic object. A boundary object shapes and is shaped by ongoing practice.

sheet described in the Canada-Care case is clearly a boundary object that was interpreted by different actors in multiple ways. But its material characteristics (it originated from an electronic document so could continuously be reviewed, reprinted and redistributed) enabled what we have called *co-development*. Co-development implies that all people (and communities) involved in the use of a boundary object are constantly engaged not simply in its use but also in efforts to modify it to meet emerging needs (and this is what the imbrication metaphor described above illustrates). For instance, the introduction of new partners in the Canada-Care network in 2014 involved a revision of the medical sheet to include new fields, acknowledging the expansion of the partnership. Similarly, a Wikipedia page is co-developed by a variety of users around the world. This plasticity is about actors' ability to change the structure of an object, on top of its content. We argued that this might explain why some boundary objects last longer than others. In relation to this chapter and to the distinction between strategic and boundary objects, it is therefore important to understand that strategic objects are generally static because they are created by change agents and represent an effort to modify the status quo (e.g., a PowerPoint presentation is created with the purpose to convince people that something needs to be done). Boundary objects, on the other hand, might have material plasticity in that their flexibility does not rest solely on the various (cognitive) interpretations that people can give to some knowledge (traditional view of boundary objects) but is also related to their ability to be modified (not only in their content but also in their structure) by the actors involved. But both types of objects are helpful (and needed) in order to innovate.

In conclusion, we can again point to the relevance of strategic and boundary objects here and that they are paramount because they 'do things'. However, this only partially explains why they are (or are not) effective. A novel stream of research, in the management literature, concerns the role of emotions and feelings that relate to the use, in practice, of objects. We briefly unpack this last issue next to provide a point of discussion and comparison.

Human engagement in objects: The role of emotions

As a final theme, we now consider the role of emotions when humans engage with objects. This is important in understanding the human-material ensembles involved in objects and their engagement and provides an interesting illustration for the purpose of this chapter. Emotions are a response to a situation perceived to be personally significant that leads to a mix of physiological changes (e.g., increased heart rate), feelings (e.g., being stressed), cognitive processes (e.g., thoughts about not being able to cope), and behavioural reactions (e.g., wanting to flee or fight). If we relate emotions to objects, one possible interpretation would be that people feel emotions (fear, happiness, anger) when they interact with an object. For instance, a manager might feel happy because of a quarter's results that are surprisingly positive, or a social media user might feel scared because someone is cyber-bullying them. These two examples can be easily conceptualised as human-material ensembles in terms of reactions to the meanings embedded in an object.

If examined using the practice perspective, emotions are not limited to feelings. Emotions are *doings*. Both strategic and boundary objects may be particularly effective if they produce an emotional as well as instrumental appeal; in the Canada-Care case, the coordinating physician's PowerPoint, the third-party documents (the literature review) and the medical sheet all underpin the attempt to improve the health of children with complex care needs – an aim that is likely to have emotional appeal when people understand the problems that the children and their families go through. Emotions are generally seen as individual reactions to a specific situation (e.g., a thief who points a gun at somebody, aims to make the victim feel afraid). However, from a practice perspective, emotions are produced relationally (Stein et al., 2014). This means that not only do relations between actors (material as well as human) produce emergent task-related outcomes (i.e., 'knowings'), they also produce emotions that can result in a particular felt quality or mood that characterises our being-in-the-world (Ciborra, 2006). Therefore, emotions are not simply experienced as an afterthought of action that is produced by an individual's interpretation of the situation, but rather emotions are a psychosocial phenomenon that emerge from collective action. Emotions are performative, themselves generating emergent outcomes (Solomon and Flores, 2003). Indeed, Dreyfus (1991) argues that a mood is always present, shaping and being shaped by our collective actions, and this mood can generate a collective energy (or its opposite – apathy, for example). This can be found in the Canada-Care case and is related, for example, to the family forum, where those present felt 'whooshed up' (Dreyfus and Kelly, 2011) by the family accounts of their struggles, leading to increased determination on their part to enact change (Newell and Marabelli, 2016). Another example refers to the sense of relief that the families experienced as the medical sheet afforded them greater flexibility to go away. The important point about this brief discussion of emotions is that objects (whether strategic or boundary objects) that produce an emotional response are more powerful change agents than objects that do not produce such a response. Of course, this can be negative as well as positive – with negative emotions produced by an object potentially leading to resistance to innovation, for example. Nevertheless, as we have identified in other chapters, resistance is not necessarily always 'bad', because it helps those leading the innovation project to identify barriers and challenges that will need to be addressed to enhance the chances that the innovation project can be successful.

Conclusions

In this chapter, we have explored the role of objects in creating knowledge and their broader role as actors within the innovation process. While traditional research views 'things' as tools that can be used (instrumentally) to share knowledge, here we propose a view that questions ways to conceive computers, systems, documents and any material actor as instrumental to human agency. We suggest that adopting a practice perspective is helpful to understand that people and objects are constantly interlocked in everyday organising practices. In this sense, this chapter draws us into considering the role of objects in the innovation process and we have seen how different types of object (strategic and boundary objects) play different roles as either tools for human actors or as agents that themselves enact change in their interaction with the social world. Considering the role of such objects is therefore very important for anyone involved in an innovation initiative – getting the right people involved is of course important to represent the different stakeholder groups and press ahead with needed changes, but getting the right objects involved is equally important. Failing to convince others of the relevance of the change (e.g., using strategic objects) and not having objects that can coordinate practices of different communities (e.g., using boundary objects) are likely to result in a failed innovation initiative. We have also reviewed how digitally produced and shared objects can have more material flexibility which potentially allows them to be co-developed over time, so sustaining their role in sharing knowledge, even when practices change.

? DISCUSSION QUESTIONS

The following discussion questions are relevant to using this chapter in teaching exercises and discussions or for revision:

- Why do we need to view material objects as actors with agency? Provide examples of digital material objects.

- What are the key differences between material and human actors during the innovation process?

- What is the difference between a strategic object and a boundary object? Why might digital objects in particular be useful boundary objects?

- How can strategic objects be used in ways that are more helpful for powerful individuals than for other stakeholders?

- What is meant by the idea that boundary objects are emergent and flexible?

- Provide an example from your own experience to illustrate the idea of imbrication.

 ## Case questions

The following case questions might also be relevant to using this chapter in teaching exercises and discussions:

1. Identify different human and material actors in the Canada-Care case and provide a description of the agency of each.

2. Which of these actors can be described as a strategic object and why do you describe it as such?

3. Which of these actors can be described as a boundary object and why do you describe it as such?

4. Why was the medical sheet such an important actor in the practice changes that were enabled in this case?

5. While the medical sheet was digitally produced, it was not initially digitally shared. What might be the advantages and disadvantages of a boundary object that is digitally produced and shared?

Additional suggested readings

Leonardi, P. M. 2011. When Flexible Routines Meet Flexible Technologies: Affordance, Constraint, and the Imbrication of Human and Material Agencies, *MIS Quarterly* (35:1), pp. 147–167.

Orlikowski, W. J. 2007. Sociomaterial Practices: Exploring Technology at Work, *Organization Studies* (28:9), pp. 1435–1448.

References

Carlile, P. R. 2002. "A Pragmatic View of Knowledge and Boundaries: Boundary Objects in New Product Development," *Organization Science* (13:4), pp. 442–455.

Ciborra, C. U. 2006. "The Mind or the Heart? It Depends on the (Definition of) Situation," *Journal of Information Technology* (21:1), pp. 129–139.

Courpasson, D. 2000. "Managerial Strategies of Domination. Power in Soft Bureaucracies," *Organization studies* (21:1), pp. 141–161.

Dreyfus, H., and Kelly, S. D. 2011. *All Things Shining: Reading the Western Classics to Find Meaning in a Secular Age*. Simon and Schuster.

Dreyfus, H. L. 1991. *Being-in-the-World: A Commentary on Heidegger's Being and Time, Division I*. MIT Press.

Hardy, C. 1996. "Understanding Power: Bringing About Strategic Change," *British Journal of Management* (7:s1), pp. S3–S16.

Jarzabkowski, P. 2005. *Strategy as Practice: An Activity Based Approach*. London: Sage.

Jarzabkowski, P., and Kaplan, S. 2015. "Strategy Tools-in-Use: A Framework for Understanding "Technologies of Rationality" in Practice," *Strategic Management Journal* (36:4), pp. 537–558.

Kaplan, S. 2011. "Strategy and Powerpoint: An Inquiry into the Epistemic Culture and Machinery of Strategy Making," *Organization Science* (22:2), pp. 320–346.

Leonardi, P. M. 2011. "When Flexible Routines Meet Flexible Technologies: Affordance, Constraint, and the Imbrication of Human and Material Agencies," *MIS Quarterly* (35:1), pp. 147–167.

Leonardi, P. M., Nardi, B. A., and Kallinikos, J. 2012. *Materiality and Organizing: Social Interaction in a Technological World*. Oxford University Press on Demand.

Marabelli, M., Newell, S., Krantz, C., and Swan, J. 2014. "Knowledge sharing and health-care coordination: the role of creation and use brokers,". *Health Systems* (3:3), pp. 185–198.

Marabelli, M., Newell, S., and Vaast, E. 2017. "Boundary Objects Survival over Time: Insights from the Field of Healthcare Coordination," in: *Academy of Management Annual Meeting*. Atlanta, GA.

Newell, S., and Marabelli, M. 2016. "Knowledge Mobilization in Healthcare Networks: The Power of Everyday Practices," in *Mobilizing Knowledge in Healthcare: Challenges for Management and Organization,* J. Swan, D. Nicolini and S. Newell (eds.). Cambridge: Oxford University Press.

Orlikowski, W. J. 2007. "Sociomaterial Practices: Exploring Technology at Work," *Organization Studies* (28:9), pp. 1435–1448.

Paroutis, S., Franco, L.A., and Papadopoulous, T. 2015. "Visual Interactions with Strategy Tools: Producing Strategic Knowledge in Workshops," *British Journal of Management* (26:S1), pp. S48–66.

Pickering, A. 1993. "The Mangle of Practice: Agency and Emergence in the Sociology of Science," *American Journal of Sociology*), pp. 559–589.

Pisano, G. P. 2015. "You Need an Innovation Strategy," *Harvard Business Review* (93:6), pp. 44–54.

Solomon, R. C., and Flores, F. 2003. *Building Trust: In Business, Politics, Relationships, and Life.* Oxford University Press.

Spee, A. P., and Jarzabkowski, P. 2009. "Strategy Tools as Boundary Objects," *Strategic Organization* (7:2), pp. 223–232.

Star, S. L., and Griesemer, J. R. 1989. "Institutional Ecology, Translations' and Boundary Objects: Amateurs and Professionals in Berkeley's Museum of Vertebrate Zoology, 1907-39," *Social Studies of Science* (19:3), pp. 387–420.

Stein, M.-K., Newell, S., Wagner, E. L., and Galliers, R. D. 2014. "Felt Quality of Sociomaterial Relations: Introducing Emotions into Sociomaterial Theorizing," *Information and Organization* (24:3), pp. 156–175.

Suchman, L. 2007. *Human-Machine Reconfigurations: Plans and Situated Actions.* Cambridge, MA: Cambridge University Press.

Wheelwright, S. C., and Clark, K. B. 1992. *Revolutionizing Product Development: Quantum Leaps in Speed, Efficiency, and Quality.* New York and Toronto: Free Press.

Yakura, E. K. 2002. "Charting Time: Timelines as Temporal Boundary Objects," *Academy of Management Journal* (45:5), pp. 956–970.

8 EXPLICIT DIGITAL CONNECTIVITY, KNOWLEDGE AND INNOVATION

Summary

This chapter considers how individuals, organisations and even regions are leveraging the digitally connected world in which we live to support innovation activities, not only in developed economies but in less developed economies as well. We illustrate how e-commerce applications generally – and communication applications specifically – can be and are being used in many different ways to support innovation in a variety of contexts. This includes using our digital connectivity to promote new forms of knowledge-sharing as well as to provide new opportunities for commercial transactions. We discuss how knowledge-sharing can now more easily take place, not just *within* organisations but also *between* organisations and individuals – a phenomenon that has been described as crowdsourcing. We also consider the problems as well as the opportunities that are emerging because of our constant connectivity.

Learning Objectives

The learning objectives for this chapter are to:

1. Understand the difference between implicit and explicit connectivity
2. Appreciate the role of digital technology *and* local actors in supporting the commercial development opportunities for communities that can help to reduce poverty

3. Recognise the benefits and problems associated with using social media within an organisation

4. Differentiate between organisationally controlled and user-controlled social media use beyond the organisational boundary

5. Appreciate the opportunities and threats associated with the peer-to-peer sharing economy

6. Reflect on the broader work–life implications of our constant, digitally supported explicit connection to work.

Case: Digital innovation in China (Taobao Villages)

Poverty still exists. This is the case in developed economies, for example in rural communities (or poor areas of rich cities), but is even more of an issue in less-developed countries. Given limited income opportunities for many in these economically poor communities, they often suffer from out-migration, with young people in particular seeking opportunities in towns and cities. Internet-enabled computing technologies have for a while been seen as a potential solution to this problem, offering rural communities access to education, healthcare and markets, through e-commerce. While in many cases the potential of e-commerce is led from the top, by governments and/or NGOs (non-governmental organisations), success will almost inevitably depend on the emergence of local leaders, and so community-driven development is increasingly promoted.

Our case considers two so-called Taobao villages in rural China that have exploited e-commerce to move many local citizens out of poverty. They are both bottom-led initiatives. These villages have used Alibaba, an online marketplace (very similar to eBay), to help increase opportunities. Alibaba developed the term Taobao Village to refer to a village where at least 10 per cent of residents are active on Alibaba and where as a whole the village generates sales of at least 10 million RMB (USD $1.6 million). Both the villages described are classed as Taobao villages.

Suichang village: Suichang is situated in the Zhejiang Province of China; before the advent of e-commerce, over 70 per cent of the 50,000 residents were farmers who had not attended high school. The average annual income of households was less than 2,500 yuan (USD $400). A local businessman returned to the village and started selling local produce as an e-tailer – setting himself up as an online retailer and selling directly to consumers. His business was successful. He later established a grass-roots Online Shop Association specifically to help others follow his success. His aim was to ensure that the local farmers were able to sell direct to their customers, using Alibaba, so dramatically increasing their returns as the distributor was removed from the value chain. He recognised the power of direct-to-consumer sales, but he also knew that it would be difficult for the low-skilled residents of Suichang to become e-tailers. The Online Shop Association therefore offered free e-commerce training to villagers, helping them to learn and make decisions about pricing and marketing strategies. This led to a rapid increase in online stores operated by villagers in Suichang and allowed many people to dramatically increase their income by directly selling local agricultural products, including bamboo shoots, tea, sweet potatoes and wild herbs.

As the number of online stores increased, people who had left the village in search of opportunities returned and started their own online stores. For instance, one man returned to Suichang and opened two online stores selling bamboo charcoal (a well-known product of this area) and very quickly increased his sales to about 40,000 yuan (USD $6,400) a month. Over time, more specialisation occurred among the villagers. The Online Shop Association helped with this by coordinating supply and demand – so that the e-tailers could essentially band together to buy goods from the farmers at a cheaper price than they would have been able to negotiate individually. This meant that the e-tailers, who initially had very little reserve and so could not buy the products they wanted to sell in bulk (and so cheaply), were able to purchase small quantities at a better price. This enabled them to make more money from the products when they sold them online. Simultaneously, the suppliers still benefited as they could sell, in bulk, to the local Online Shop Association (rather than to the traditional large distributor) and could work with this Association to plan what produce to grow based on a better understanding of demand.

Beishan village: Beishan is situated in the province of Jinyun. Prior to the move to online selling, the village was dominated by agricultural work, with 92 per cent of the workforce employed in this industry. As with Suichang, annual incomes were very low. In this village, e-commerce was initially stimulated by a local clay-oven bread producer (Mr Lv) who learnt about the opportunities of online selling from a friend. He saved enough money to start an online store with his brother, buying outdoor equipment from the wholesale market and selling online direct to customers for a mark-up. Initially, sales were very slow but with good customer service his business grew, as did his income. His success was soon visible since he was able to buy a BMW car. His neighbours then asked about his success and he started to help others develop their own online stores, largely following his model of focusing on selling outdoor equipment, including backpacks, sleeping bags and tents. One villager started selling online, following the example of Mr Lv, and while at first he indicated that business was tough, he persisted. Within a couple of years, he had an annual turnover of 10 million yuan (USD $1.6 million). In this case, again villagers who wanted to set up an online store had very limited reserves so, initially, Mr Lv, now a successful entrepreneur, allowed them to buy products from his stock only once they had orders, at only just above the wholesale price he had paid. This meant they did not initially have to buy stock that would tie up their very limited capital. Later, Mr Lv created his own outdoor activity brand – BSWolf – and he then sold this to local villagers to sell online so that he became a supplier of products as well as an e-tailer.

Evaluating the two villages

In both these examples, an individual begins to experiment with e-commerce, working from home using digital technologies to sell directly to consumers, either local products or products bought from wholesalers that could be sold at a mark-up. Others observe the success and come and ask for help to do the same. Gradually, an entrepreneurial climate is established in the village as more and more people start to undertake e-commerce, either directly setting up online stores or providing services to the e-retailers, such as logistics, graphic design, packaging and online help-desk services. There were many examples in both villages where individuals who had left the village to seek opportunities elsewhere returned to provide workers for the burgeoning service industry that was needed given all the online retail activity. In this way, both villages used e-commerce to help alleviate poverty among residents while attracting new people to the village (some who had previously left, others who saw opportunities) so that the local economy could grow even more.

In both villages, local residents were central in helping villagers exploit digital technology to sell products directly to consumers, cutting out the middleman as a result. This is referred to as disintermediation in the distribution channel. In Table 8.1, we identify the crucial actors that seem to be central for this kind of digitally led development to be successful.

In this chapter (and also in Chapter 9), we consider ways in which new developments in digital technology are changing the innovation process. This case helps to illustrate and discuss key topics and themes here.

Introduction

Here, we explore new developments in digital technology and how they are changing the innovation process. This includes new opportunities for commercial transactions and new ways to source knowledge for innovation (e.g., from 'the crowd' rather than relying simply on employees or partner organisations). We have already introduced the concept of open innovation in Chapter 6 and identified different forms of open innovation, including using platforms and competitions. We look in more detail at this in this chapter under the heading of *crowdsourcing*. Even more fundamentally, our constant digitally based connectivity is changing the way that we think about knowledge

Table 8.1 Critical actors in the rural e-commerce ecosystem

Actors	Roles and examples from the case
Grassroots leaders	Villagers or grassroots organisations that initiate, lead and shape the development of an ecosystem. They are key actors who provide initial support for the emergence of e-tailers (e.g., training, product supplies).
E-tailers	Villagers who sell products though e-commerce.
Third-party e-commerce service providers	Villagers who provide services to support e-commerce operations in villages, making it easier for e-tailers to do business. Services include logistics and delivery, product packaging, marketing services, website and graphic design, photography and online customer service outsourcing.
E-supply chain partners	Villagers who produce, supply or distribute products that are sold via e-commerce.
Digital platform sponsor	Providers of the technological infrastructure for the e-commerce, such as product display, searching, transaction processing, payment, reports etc. In this case, Alibaba.
Institutional supporters	Institutional stakeholders such as government and telecommunication providers who play a functional role in improving such infrastructures as road transport and telecommunication services and a symbolic role in providing legitimacy for entrepreneurial risk-taking.
Online consumers	Buyers, including those from urban areas, who purchase products offered by the villages through e-commerce.

and how knowledge is used for making decisions – with, for example, big data privileging algorithmic decision-making rather than conventional human expert judgment (we discuss this in Chapter 9). In other words, in these two chapters (8 and 9), we focus on the increasing presence of digital technologies that invade all aspects of our life – both work and play. We consider our *explicit* use of these new technologies to share knowledge, collaborate and engage in transactions (the focus of this chapter) as well as the *implicit* collection of digital data about everything that we do that can then be mined as a basis for making decisions related to innovation (the focus of Chapter 9).

As we have seen, the ability to draw distinctions from patterns in data and so elucidate meaning in a particular context has traditionally been considered the essence of knowledge and the basis for decisions. Different approaches to knowledge management afford different opportunities and challenges in relation to this ability. Therefore, in the academic literature, a distinction is drawn between *repository* (codification) and *network* (personalisation, or peer-to-peer) approaches (Hansen et al., 1999; Alavi and Leidner, 2001; Newell and Marabelli, 2014) to knowledge management, with either or both being valuable, depending on the purpose and type of knowledge involved. Specifically, the repository approach can be useful for managing more *explicit knowledge*, whereas the network approach can be useful for managing *tacit knowledge*. A repository provides access to what others have done previously and this can be useful as a basis for making decisions in a current context. Networks provide forums for discussions and access to experts in other places that can be used to help solve current problems and identify potential solutions; for example, where an innovation project needs to make some kind of decision and requires additional input in order to do this.

Both the repository and network approaches are aimed at improving knowledge processes *within* an organisation, based on the assumption that 1) a firm's competitive advantage derives from the unique mix of knowledge and skills that reside within its boundaries and that 2) organisations exist precisely because it is easier to manage knowledge, especially knowledge for innovation, within the organisational structure. The latter finds form in the so-called knowledge-based view of the firm (Kogut and Zander, 1996; Nahapiet and Ghoshal, 1998). Therefore, while a firm needs to be able to *absorb* knowledge from beyond its boundaries (Cohen and Levinthal, 1990; see Chapter 2), it must also *protect* its knowledge since this is traditionally viewed as a key source of advantage (Trkman and Desouza, 2012; see Chapter 6). Recently, however, a number of fundamental IT changes have afforded new ways for organisations to explore and exploit knowledge for innovation (Durcikova et al., 2011), opening up opportunities for new players to be involved. Simultaneously, these new digital developments raise some concerns related to where knowledge is created, by whom and what are the underpinning practices that enable knowledge-sharing across boundaries (see Chapter 6) such as between different professional communities that access the same online resources.

In this chapter, we first consider digital connectivity in its explicit and implicit forms. We then focus on explicit connectivity. We provide examples of e-commerce innovation in less-developed countries and then consider examples of social software and social media and their role in innovation processes. We end with a consideration of the crowdsourcing phenomenon in all its guises and with some reflection on the sharing economy and the broader impact of our IT connectivity on our work–life balance.

Implicit versus explicit digital connectivity

One relevant feature of today's workplace, as well as increasingly in our homes and leisure spaces, is that computers, including mobile computing devices (Sørensen and Landau, 2015), are everywhere. This has been termed *ubiquitous computing* (Satyanarayanan, 2001) and means that we are connected to other people and organisations either explicitly or implicitly. We therefore distinguish between implicit and explicit connectivity. *Explicit connectivity* includes, for instance, when we use Facebook to share our thoughts with 'friends' or Twitter to communicate with a wider audience or when we use an online shopping channel, such as Alibaba in the Taobao Village case above, to sell and purchase a product. *Implicit connectivity* includes, for example, when our smartphone records where we are and our whereabouts can be tracked, when a cookie records that we have visited an online shopping site and searched for a particular product or when we wear a fitness monitor that tracks our activities (Newell and Marabelli, 2014) (Figure 8.1).

The distinction between explicit and implicit connectivity is important in relation to thinking about new ways in which digital technologies offer opportunities for exploring and exploiting knowledge for innovation. Explicit connectivity underpins the emergence of Web 2.0 (or Enterprise 2.0), where applications have been developed to allow communication and transactions, typically based on applications that connect to the internet. In this context, people (including strangers) are able to connect and share

Figure 8.1

Smartphone apps allow the sharing of our whereabouts constantly
©Getty Images

knowledge and transact with one another based on some kind of mutual interest or objective (McAfee, 2006). The common term for the applications that support this sharing is *social software* and, especially when applied in an organisational context, the stated purpose of using such social software is to take advantage of many people's ideas and expertise in order to either identify new opportunities or solve problems through collaborative working (Huang et al., 2013; Baptista et al., 2017). Importantly, those involved may not just be employees or those working in partner organisations; rather, they can be anyone, anywhere, as with the phenomenon of crowdsourcing (or crowd involvement), which we will consider later in this chapter. In relation to transacting with others, there are various kinds of online applications that allow such transactions, including sites like Alibaba and eBay but also increasingly sites like Airbnb and Uber that we discuss later (Figure 8.2).

Implicit connectivity is different because it does not rely on human actors purposefully inputting their ideas and expertise in order to contribute to innovation processes. Instead, objects (things we use in our everyday practices) are increasingly digitised and used to capture what is being done, by whom and/or where, through sensors or tracking devices. This, then, creates big data that can be analysed to identify, for example, connections that may provide ideas for new products or services or to monitor individual behaviour for security innovations. However, these types of innovation can also have unintended social consequences, such as invasions of privacy, and these issues will be considered in Chapter 9.

 Key Concepts: Connectivity

Explicit: Use of internet-connected devices (computers but all kinds of other mobile technologies) to accomplish a goal (e.g., posting a photo on Instagram or selling a product on Alibaba or eBay)

Implicit: Use of digital devices (computing devices but also other internet-connected technologies) so that activities are tracked and data are then used for a purpose not linked to the original intention of use (e.g., online purchases or browsing activity tracked and sold to advertisers who can then better target marketing information)

Explicit digital connectivity and e-Commerce

Organisations can be connected to their various stakeholders through virtual, online communications, and these connections can be exploited for innovation. This is the case in developing economies where access to computing technology and the internet is less widespread but where mobile technologies are becoming more widespread such as in India (Abraham, 2006).

Figure 8.2

eBay and similar websites connect people from all over the world
©PhotoDisc/Getty Images

Indeed, improving access to internet-connected computing devices, and so reducing the so-called *digital divide*, is seen as a key goal for enabling developing countries to become more equal partners in the global economy (e.g., Skaletsky et al., 2016).

The Taobao Village case introduced at the start of this chapter provides an example of how a developing economy can exploit innovation opportunities from the increasing access to internet-connected computers and mobile devices to improve economic conditions for poor communities. Such examples, which can be generically described as e-commerce innovations, are of course common in developed economies as well. Indeed, it is very difficult to imagine a business being successful were it not to have some kind of internet presence that allowed stakeholder interaction by, for instance, directly reviewing and purchasing products and services online (Figure 8.3).

Figure 8.3

e-commerce platform
©PhotoDisc/Getty Images

The opening case describes the power of digital infrastructures (here, the internet) to enable economic development even in rural locations where there may be limited opportunities to access resources, knowledge and markets. It also showcases the intermingling of the social and the technical that we discussed in Chapter 7. In this respect, it is worth noting that the Taobao Village cases illustrate that it is necessary but not sufficient that businesses have access to the internet. They also need the encouragement and support of local entrepreneurs who help them understand how to exploit digital technology for commercial success. However, perhaps more importantly, the two cases show different patterns in relation to how this type of bottom-up digital technology exploitation for development actually occurs, showing that there can be different routes to such e-commerce innovation. Therefore, the two villages each took a different route in terms of how the different local actors emerged and how they worked together in order to develop the capabilities necessary for successful e-commerce-enabled innovation.

Based on this, two different approaches to rural e-commerce innovation have been identified: *orchestrated* and *organic* (e.g., Cui et al., 2017). In the *orchestrated* approach, grassroots leaders (individuals who were born locally but who then moved away to become successful business leaders and later return to their home village in a bid to stimulate the local economy) recognise the limited initial capabilities of the villagers (e.g., their lack of digital expertise) and take action to remedy the shortcomings. Thus, in Suichang village, leaders created an Online Shop Association that could support others in their e-commerce innovations by, for example, offering digital training to help villagers start their own online retail business or providing marketing advice. The grassroots Association also helped to coordinate supply and demand so that local e-tailers could have greater bargaining power because they could pool the stock that they bought to sell on. The association also organised online sales campaigns and showcased successful examples of e-retailing so that others could learn from their experiences.

The alternative approach is *organic* in the sense that no central coordinating association is established, but rather there are multiple people involved in a general process of discovery and learning. In our case, initially one family was successful in starting an e-commerce retail business and others observed the success and then asked for advice on how to do this. The successful e-commerce entrepreneur helped fellow villagers by agreeing to supply them with stock to sell on their online business sites as they made sales. They did not have to pay for stock in advance, which they could not afford to do at first, having very little access to financial reserves. But more generally in this case, the success of one family in starting a successful e-commerce business provided a role model for others to copy, showing them that it was possible to be successful, and as a result this gave others the confidence to try themselves and persevere over time. Table 8.2 synthesises the two approaches described.

This example is ultimately useful in helping to understand how e-commerce innovations can promote economic development in less-advantaged areas – whether in developing countries or rural or deprived parts of developed countries. We can conclude, therefore, that explicit connectivity might help overcome socioeconomic disadvantage, with centrally managed digital providers (e.g., Alibaba or eBay) giving people in rural and deprived areas broader access to a market for selling products and services (as well as providing access to goods and services for people in these same communities that were previously not available). However, e-commerce is not the only way to pursue innovation with explicit connectivity. Other forms of software application, such as social media applications and, more generally, user-generated content systems, offer opportunities to create and share knowledge, therefore to innovate, as we articulate next.

Explicit digital connectivity and use of social media within organisations

e-commerce rests on using our explicit connectivity to enable transactions between people and organisations. This can obviously provide new opportunities for innovations, including at the community level rather than simply the individual or organisational levels. However, on top of this transactional affordance of digital technology, there are also communication affordances

Table 8.2 Orchestrated and organic approaches (from Cui et al., 2017)

Orchestrated approach	Association organises training and workshops and Association mobilises the villagers	Association provides an operational infrastructure that coordinates the supply and demand	Association provides support to ensure the villagers Association showcases successful e-tailers
Organic approach	Pioneer shows by doing; self-learning by villagers; Pioneer co-evolves with the villagers	An operational infrastructure of distributor-agent model emerges from the co-evolution of the ecosystem actors	Success of the pioneering e-tailer inspires the villagers Mutual support through peer sharing and coalition building

that we need to consider, especially in relation to the use of social software to enable virtual knowledge-sharing among many, distributed people (Yuan et al., 2013). This social software-enabled knowledge-sharing either can be controlled by the organisation – where the organisation seeks out knowledge input from its distributed employee/partner network or from the broader crowd – or can be out of the organisation's control – where employees or those external to the organisation (e.g., users of its products or services) generate content about an organisation and its products and services or where users connect to each other to offer goods and services.

A number of recent studies have focused on the positive impact of a specific type of social software used within organisations, known as Enterprise Social Media (ESM) systems. Leonardi et al. (2013) provide a comprehensive review. This type of social software can open up new opportunities for increased communication and knowledge-sharing within an organisation and so could foster increased employee participation and engagement (Kaplan and Haenlein, 2010; Skågeby, 2010; Majchrzak et al., 2013) and could in turn lead to more people being involved in innovation projects and in organisational strategy-making (Huang et al., 2013; Baptista et al., 2017). Of course, implementing a social media platform within an organisation (such as Yammer) will not automatically lead to employees participating in sharing content with others or in increased collaboration. Prior studies have therefore considered how to encourage more participation on social media sites within organisations, noting that there can be a reluctance among employees to voice their views, especially when they are at odds with those in more senior positions or when they want to criticise their bosses (Treem and Leonardi, 2013; Kuegler et al., 2015). Where a platform is imposed by managers (Choudrie and Zamani, 2016) then, what is shared may be quite restricted. However, even when successfully implemented, ESM are not without potential negative impacts (Champoux et al., 2012; Hildebrand et al., 2013). For example, social media can lead to a more open internal communication environment that allows diverse views to be shared. This can be a good thing of course but can also lead to conflicts between people with different ideas. This more diverse exchange of views can detract from the unified message that perhaps those at the top traditionally like to communicate and control (Huang et al., 2013). In Chapter 3, we saw how CommCo had overcome this problem by developing an ambidextrous approach that allowed both a central communication channel for top-down messages and a peer-to-peer channel.

The above highlights that innovation can be pursued in various ways and that our increased explicit digitally based connectivity can promote innovation but that one always has to consider the technology and the social system together to understand the impact of digital innovation. Thus, it is hard to develop a 'recipe' (or a best practice) for how to use social media to make innovation processes effective, as many solutions have pros and cons and it is hard to predict what specific solution (for example, the balance between organisationally controlled content and peer-to-peer content) should be adopted in an organisation.

 Key Concepts: Digital Innovation

Digital divide:
A socioeconomic divide (generally considered at the country level) arising from the different availability of digital infrastructures. Studies suggest that the availability of digital technology speeds up socioeconomic progress, therefore leaving behind countries where such infrastructures are not present. In other words, the unequal diffusion of digital technology might increase the socioeconomic gap between haves and have-nots, while the adoption of digital technology might help to decrease this gap

e-commerce:
The buying and selling of goods and services over an electronic network, primarily the internet; as a result, e-retailers can sell products direct to customers, cutting out the intermediaries so that they can maintain a greater share of the profits as well as interact over distance

> **Enterprise social media (ESM):**
> Computer-mediated technologies that facilitate the creation and sharing of knowledge, information, ideas, career interests and other forms of expression via virtual communities and networks used within an enterprise

Explicit digital connectivity and crowdsourcing

In addition to the various organisationally (more or less) controlled or managed social software that is used 'within the organisation' (i.e., ESM), we next describe how social software can be used to involve external actors. Crowdsourcing (where the source/content of social software originates outside the organisation's boundaries) can be organisationally controlled and user-controlled, as we discuss next.

Organisationally controlled crowd involvement

Since the launch of the internet, innovative business models have been developed that have created new forms of business (e.g., selling books or clothes on the internet rather than in physical shops). Indeed, internet shopping is now so common that we do not think of it as a new business model because in many industries it is imperative for a firm to sell on the internet either instead of or alongside traditional retail outlets; its very survival may depend on it (Kim et al., 2007). Additionally, e-commerce has enabled not just new business models but also new opportunities for whole regions, which are now able to compete where before they were perhaps cut off by their location or their limited access to resources. We saw this in the opening case in relation to rural Taobao villages in China.

Aside from these efforts in e-commerce innovation, organisations are now beginning to think about using digital technology to involve a much broader range of people in the actual processes of developing innovations rather than simply seeing technology as a conduit to source or sell products and services (Greer and Lei, 2012). This is all part of the move by organisations to make their innovation processes more open, as discussed in Chapter 6. It has long been recognised that customers are an important source of knowledge for organisational innovations, and von Hippel (1978) coined the term *customer-led innovation* to describe this. With the widespread availability of the internet, connecting to customers and other stakeholders has become much easier and more prevalent. This is where social software comes into its own, being used on the assumption that, even though there may be different views and ideas, in general, 'the crowd knows best' (Boudreau and Lakhani, 2013). This is related to the idea sometimes used in quiz shows that allows a participant to 'ask the audience' if they are not sure about an answer;

the audience may be divided in their advice, but the assumption is that the majority will be right more often than not.

We can also use the crowd to find answers to problems that employees are not able to solve, because the crowd has more diverse knowledge that can be tapped into – especially if as many people as possible are enticed to get involved. We can use the 'wisdom of the crowd' to enable fast and effective open innovation (Chesbrough and Garman, 2009), by for example, using online contests to stimulate contributions (Boudreau and Lakhani, 2013). This type of crowdsourcing, in other words, is based on the idea that knowledge diversity can generate innovation more effectively than uniformity (Majchrzak and Malhotra, 2013), with a number of studies across different problem areas demonstrating this advantage (e.g., Boudreau and Lakhani, 2013), and that digital applications can afford the bringing together of this diversity. In particular, by engaging the crowd, companies have access to users' knowledge about product design and use, which can supplement in-house knowledge. In this way, extra- (rather than intra-) organisational knowledge processes come to the surface. An example of a company that uses a digital platform (in this case, a platform built in partnership with platform provider Chaordix) to connect with its customers to generate and test ideas for new products is Lego's Ideas site (Shlagwein and Bjorn-Andersen, 2014). This allows people to submit ideas but also build on others' ideas. Once an idea for a new product has received many endorsements from the Lego community, it then gets reviewed within the company and, if selected, the product gets developed and commercialised by the company, with the community member(s) who generated the idea receiving some of the profits.

Crowdsourcing successfully

There are numerous examples of other digitally facilitated crowdsourcing success stories. Such examples principally relate to the use of crowdsourcing to generate ideas for introducing new products and services. For example, some sites simply ask for customer input (e.g., MyStarbucksIdea.com, which solicits ideas from customers about how to improve its service or suggest new products to introduce), whereas other sites offer prizes for the best solution (e.g., vitaminwater.com, which asked for customer input to select the next flavour for its water but also offered a prize for the best design of the bottle label). As well as such sites that generate ideas for innovation, other sites provide the opportunity to use the crowd to deliver an 'instant workforce'. In relation to the latter, the best-known example is Amazon's Mechanical Turk (MTurk), which provides 'an on-demand scalable workforce' where people can post jobs that others can sign up to do for an agreed rate of (usually very minimal) pay. MTurk has been especially popular among academic researchers as it allows them to conduct research projects by posting for volunteers on the site. This potentially provides a reservoir of survey participants and avoids a common problem relating to small sample size for many social science researchers. The use of MTurk is not without its issues, however, especially for the crowd-workers themselves. A study by Deng et al. (2016) highlights the duality of the freedoms associated with crowd-working (such as the ability to take on work when and where it suits) and, at the same time, issues associated with poor levels of remuneration and scams (i.e., non-payment arising from spurious rejection of work done by job requesters). In addition, ethical issues have emerged that are associated with potential reidentification of crowd-workers, the individuals who are warranted anonymity upon the agreement to participate in a study (Marabelli and Markus, 2017) (Figure 8.4).

Aside from these more commercial examples of using social software to generate input from the crowd (whether in terms of ideas or simply a labour

Figure 8.4

It is often difficult to find the right balance between using technology intensively and responsibly

©GETTY

force), there are crowd sites that have a more socially motivated objective. OpenIDEO is one of the best-known examples of this, setting challenges or establishing programmes for collaborators to work together to identify solutions for tough or 'wicked' (problems that seem to be intractable) social problems such as tackling climate change or improving education for refugees around the world.

knowledge-sharing in crowd activity

Research has focused on how to maximise knowledge-sharing in collaborative crowd forums (Faraj et al., 2011), especially in contexts where there may be no immediate gain to those sharing their knowledge and despite potential free-riding of some (e.g., Whelan, 2007; Davison et al., 2013). While there has been focus on increasing contributions from the crowd, another finding is that most crowd platforms produce only independent contributions and do not encourage co-creation (i.e., others commenting on posted ideas and feeding from each other to stimulate more innovative solutions through 'creative abrasion' – Leonard-Barton, 1995). This can lead to the situation that, while many ideas are posted, these ideas are not further elaborated upon (Majchrzak and Malhotra, 2013), and this restricts knowledge exploration. Moreover, Bayus (2013) looked at Dell's IdeaStorm and found that those who contribute good ideas through crowdsourcing are those who generate multiple ideas rather than single ideas. However, he also found that, once an *ideator* had had an idea implemented, they then found it difficult to come up with further valuable ideas for the organisation, because their subsequent contributions tended to be very similar to those already implemented.

The above studies suggest that we need to consider how to design social software platforms that best afford crowd participation and the co-creation of knowledge. In this vein, Majchrzak and Malhotra (2013) suggest that co-creation is restricted by three tensions in existing crowdsourcing architectures: 1) the simultaneous encouragement of competition and collaboration; 2) idea evolution takes time but crowd members spend little time; and 3) creative abrasion requires familiarity with collaborators yet crowds consist of strangers. In response to these tensions, they propose architectures that might better afford the co-creation that is needed for successful crowd-supported innovation. For instance, they suggest designing systems that encourage idea *evolution* rather than simply idea generation and incentivising contributors who work on others' ideas (i.e., rather than rewarding just the winners of a competition). In this way, digital technology is not just an enabler of crowdsourcing but can be designed to shape and optimise open innovation afforded by social software (Majchrzak and Malhotra, 2013) while taking into account the ethical issues raised by Deng et al. (2016).

While these crowd applications and research to show how to maximise online involvement are useful, organisations also need to think carefully about when they want to try to include the crowd. As von Krogh (2012,

p. 154) argues, while social software offers considerable promise for knowledge management, it also 'raises fundamental questions about the very essence and value of firm knowledge, the possibility of knowledge protection, firm boundaries, and the sources of competitive advantage'. Rather than knowledge being a precious resource to be shared within the organisation to create value (but protected from the outside to achieve competitive advantage), social software and the associated idea of crowdsourcing potentially turn this traditional mantra on its head. Utilising social software for open innovation, therefore, needs careful consideration, particularly in relation to how and what knowledge to protect and what to share with others outside the firm.

While this organisationally controlled type of crowdsourcing is perhaps most obviously related to a firm's innovation strategy, user-controlled crowdsourcing can also be relevant, as we discuss next.

User-controlled crowd involvement

Many digital sites allow individuals to comment on an organisation's products and services (e.g., in the form of anonymous reviews or comments on sites like Amazon and TripAdvisor). Such reviews have been found to influence what stakeholders (and, in particular, customers or potential customers) feel about a firm's products or services more than a firm's own market-generated content that attempts only to educate customers about the benefits of their products and services (Goh et al., 2013). Early literature focusing on how far crowd reviews on social software platforms influence purchase behaviours was inconsistent. However, more recent studies seem to agree that online reviews do affect organisational performance and are generally perceived as a threat because they can influence external reputation (Orlikowski and Scott, 2014). Some organisations try to overcome this threat by manipulating reviews in order to improve ratings (e.g., on sites such as TripAdvisor). Competitor organisations have also been known to post bad reviews of their rivals in an attempt to achieve some competitive advantage (Figure 8.5).

Ethical issues are thus prevalent because, while anonymity might make users feel more comfortable in expressing their opinion freely, this can also lead to individuals abusing others given that they don't have to disclose their identity. Indeed, Hu et al. (2011) conducted an algorithm-based text analysis on over 600,000 online reviews related to 4,490 books focusing on how the writing style of reviews changes over time. They found that 10.3 per cent of the books in their sample had been subject to online review manipulation. As an example of this problem, the *Sunday Times* (Henry 2015) reported on how journalists had created an eBook entitled *Everything Bonsai!* that was ghostwritten to be purposefully bad – full of errors and written over a weekend. They then arranged for professional review writers to each write five reviews of the book, spending a total of £56, at £3 per review. These fake

Figure 8.5

Millions of people daily post online reviews to various websites
©Getty Images/EyeEm

reviews led to the book getting to the top of the Gardening and Horticulture bestsellers category on Amazon UK's kindle site. Online retailers recognise this problem, however, and are trying to identify and prosecute such fake reviewers, as explored in an article in the *Guardian* (Gani, 2015) entitled 'Amazon sues 1,000 fake reviewers', since they essentially undermine the business model upon which online-only retailers rely and so can distort the success of new products or services (Figure 8.6).

Peer-to-Peer crowd involvement: The sharing economy

Aside from the examples explored above regarding how social software is changing the relationships between organisations and their stakeholders, social media are opening up opportunities for peer-to-peer innovations in products and, especially, services. These initiatives go under the much broader umbrella term the 'sharing economy'. The sharing economy stems from online movements such as the 'open source initiative' at opensource. org, which provides many different software programs along with their source code, so programmers can make modifications – generally improvements – and register the software under the GNU operating system's GPL

Figure 8.6

We cannot ignore the dark side of innovation uses
©Getty Images/Stocktrek Images

(general public licence), so the software is also free. This sharing economy initiative allowed the development of the Linux operating system. Linux today represents one of the most widely used internet software platforms for web and mail servers, believed to provide greater security and higher performance than its rival, Microsoft (and Linux software can be used at no cost). The open-source and GNU-licenced software movements are examples of the early stage (1990s) of sharing economy initiatives.

Nowadays, the *sharing economy* is a term used more broadly to identify any transactions that are undertaken via 'online marketplaces', so not simply 'peer-to-peer' free transactions (as with the Linux movement) but also peer-to-peer money-based transactions. Some well-known examples include Airbnb, which allows home owners to rent out their home (room, apartment, house) to people who may otherwise have used a hotel or similar, and Uber, which allows car owners to offer taxi services to those looking for rides and who may otherwise have used a licensed taxi company. Critics argue that these innovations in the digital sharing economy are leading to a reduction in well-paid secure jobs as more people undertake this form of work for low wages and with no job security or benefits. However, advocates argue that it is opening up opportunities for small entrepreneurs to start their own business and work more flexible hours that might suit their other responsibilities and interests. Advocates also argue that it is not simply about taking customers away from established businesses but that it is rather to do with increasing demand for the particular type of service because that service is now offered at a lower cost. The argument is, ultimately, that more people take rides via Uber than would have taken a taxi or use Airbnb than would have gone to visit somewhere and stayed in a hotel. Of course, these opportunities associated with the sharing economy are conditioned upon the availability of reliable and accessible infrastructures (e.g., broadband connection offered at an affordable price, see Gulati et al. (2015)) (Figure 8.7).

Those who believe that these peer-to-peer sites are exploitative are using the legal system to try to prevent the expansion of this type of innovation. Thus, there are now legal battles in different countries and cities (see the 2016 essay by Juliet Schor entitled 'Debating the Sharing Economy', which looks at how individuals might regain bargaining power even if involved in a for-profit peer-to-peer service). While for traditional companies such peer-to-peer sites can be a threat that some are trying to fight, other companies are recognising the power of such a sharing economy and are getting involved themselves (e.g., Avis, the car rental firm, paid USD $500 million for the car-sharing service, Zipcar). Furthermore, while initially some cities prosecuted home owners who took in short-term tenants through Airbnb (because they did not have planning permission to be a short-term holiday let or because they did not pay hotel tax), other cities are recognising the potential value of this type of innovation for the local economy and are moving to

Figure 8.7

The sharing economy has reshaped many industries worldwide, the most famous probably being the taxi industry

©Getty Images/Tetra images RF

legalise such activity. So, for example, the law has been changed in the UK in that homeowners can let out their home for up to three months per year without registering as a hotel.

There are other downsides to this type of digital innovation. For instance, Edelman and Luca (2014) expose the problem of digital discrimination by showing how, on Airbnb, having those selling their products and services disclose personal details, in order to build trust, also has the potential to reinforce existing forms of discrimination. They found that white hosts were charged 12 per cent less than hosts of colour for the same quality of housing. Ultimately, then, while these innovative forms of peer-to-peer sharing are potentially advantageous for some consumers (enabling them access to cheaper and more varied service options), they are not without their problems for providers, consumers or indeed local communities.

 Key Concepts: Crowd Involvement

Organisationally controlled crowd involvement:
Using explicit digital connectivity to involve people outside an organisation's boundaries to provide ideas and suggestions for innovations

User-controlled crowd involvement:
Use of explicit digital connectivity by people outside an organisation's boundaries to provide feedback and suggestions to an organisation that can help with innovation

Peer-to-peer crowd involvement:
Using explicit digital connectivity to innovate in the provision of products or services by facilitating developers and users or buyers and sellers to transact with each other directly in what is described as the sharing economy

Explicit digital connectivity and work–life boundaries

All the previous examples of how our explicit digital connectivity influences innovation processes are underpinned by the fact that so many of us now spend a large amount of time on digital devices, using social software and other applications to conduct our work and/or leisure activities. We live in a connected world. Above, we have focused on how constant explicit connectivity has many potential opportunities and challenges for regions, business organisations and individual entrepreneurs. It is also worth considering the impact of constant explicit connectivity on individuals and their personal lives because if people (as employees, crowdsourcing participants or

Figure 8.8

Working from home is more and more common, yet it also poses social issues
©Getty Images/iStockphoto

consumers) are impacted negatively, it may have a consequence for the success of a new product or service. In particular, the fact that we can now remain constantly connected to our work (and indeed more and more people are working from home) raises the issue of how this is influencing our work–life balance (Stein et al., 2015). In this respect, recent research suggests that the boundaries between work and life are being re-negotiated, so that actually describing a physical and temporal work–life boundary is not particularly relevant for many anymore (Figure 8.8).

When someone is sitting watching TV with their children at home while doing work emails, it is clear that there is no longer a clear boundary between work and home life. Professor Jon Little from Lancaster University led a major project looking at how social media and constant explicit connectivity is influencing how we work and play (Digital Brain Switch, 2015). The study suggests that we need to rethink the idea of work–life boundaries. It suggests a number of ways in which work–life boundaries are being re-negotiated.

The changing nature of work–life boundaries

The above considerations should be central when considering explicit digital connectivity and innovation. The changing nature of work–life boundaries includes the following:

1. ***Blurring boundaries between work and home*** – Rather than a clear physical and temporal distinction between work and home life, the research identified a blurring of the boundaries. For instance, a person would sit and do work emails, read and add to a blog and then do some online shopping and then log off and go back to non-virtual activities, whether work or play. This blurring of the boundary between home and work manifested in a number of new issues being confronted by individuals.

2. ***Emotional spillover*** – The blurring of boundaries makes it easier for emotions to spill over between home and work. For example if a person is at home but doing some work emails in the evening and gets an email that is annoying or frustrating, then this can easily negatively affect the mood of his or her partner, even if the bad mood has nothing to do with them. Of course, this emotional spillover is always possible from work to home (and vice versa), but with the temporal and physical blurring of work–home boundaries, this spillover (which can also of course be positive) is much more difficult to avoid.

3. ***Constantly replanning our time*** – Our constant connection means that others at work (or at home) can expect or at least ask us to do things that regularly require us to change our plans. In the past, I could have been phoned at home to add a meeting to my diary for the next day, requiring me to get to work earlier than planned. However, this was seldom done. Now, emails can be sent and read in 'out-of-office' hours, meaning that people have to be more flexible and change their plans to accommodate the new demands that they are made aware of much more frequently in the online environment.

4. ***Managing public and private online identities simultaneously*** – Given that work and non-work activities both take place online and often almost simultaneously, it can be difficult to manage the separation between our public and private identities. People handle this in different ways: for example, having different email accounts for work and home or using different types of social media (e.g., Facebook for private posts and Twitter for public posts). Many experience tensions in this respect, however, struggling with how much of their 'private' self to reveal online. Another study (Marabelli et al., 2016) identified that there are demographic differences in this respect, with younger people being less concerned with a merging of their public and private selves, while older people experienced this more as a tension, handling this tension by not revealing much of their private life online.

5. ***Getting 'sucked in' by digital technologies*** – The issue of the need to replan constantly is related to another issue identified by the research: that many of us find it very difficult to switch off from our work because of the ease of 'doing just one more email'. Moreover, since we may have

done some online shopping during our work hours or read an online newspaper, it is easy to justify this, helping us to potentially cope with the blurring between work and home life in the online context.

6. *Maintaining our digital lives* – As well as the blurred boundaries between home and work, the study also identified the time that it takes us to continue our explicit connectivity. What we can learn from this is that actually keeping our digital lives manageable involves considerable maintenance work – the researchers called this 'digi-housekeeping'. Digi-housekeeping includes dealing with emails and ensuring they are organised or deleted, sorting files so that they are accessible (often to others as well, if using shared folders), deleting posts that are no longer relevant and so on. This aspect of our online lives often takes more time than is acknowledged but can become very self-evident after a period away when we can feel overwhelmed by the amount of emails, for example, that we need to deal with on our return.

7. *Physical inactivity* – With the use of mobile technologies and wireless connections, it is not simply that we can work at home as well as in the office; rather, we can work anywhere – in the bathroom or in the garden as well as at our home desk. This erodes physical boundaries as already discussed. But another consequence of the increasing amount of time spent in online explicit connections is that most of this activity involves us sitting and being glued to the computer screen. The lack of physical activity involved in much modern-day work – even talking to a colleague now typically involves online chat or email – has wider repercussions for the health of individuals, as evidenced by the rise of diabetes, for example.

Key Concepts: Issues Associated with Digital Innovation

Work–life balance:
The balance and boundary between work and the rest of our lives is being changed by the potential of our constant explicit digital connection to work wherever we are and whatever time of day

Work–life balance and digital innovation:
There are myriad ways in which digital innovations are making the balance between work and the rest of life more blurred, including the need to manage multiple identities, emotional spillover between work and other life, the time spent in digital maintenance and the constant interruption of digital communication

Conclusions

The digitally connected world in which we now live offers many different opportunities for innovations at the regional, organisational and individual levels. There are some very positive examples of how this connectivity can help reduce poverty in remote communities, significantly improving standards of living by providing access to previously unavailable markets through online transactions. There are also some excellent examples of organisations using this explicit digital connectivity to involve more people in designing new products or services or even solving some major 'wicked' problems that still exist in our world today. And at the individual level, e-commerce platforms and the sharing economy provide potential entrepreneurial opportunities to start new businesses or simply help others out when we have knowledge or resources they need. At the same time, there are also major challenges that are exposed by this increased global digital connectivity – challenges that again are manifest at regional, organisational and individual levels. There are no easy solutions to these tensions, but understanding and discussion can help us think about, for example, how we might want to find the balance between our work and the rest of our lives or between the time we spend on our digitally enabled communications and the time we spend in face-to-face communications. And as managers we also need to think carefully about how we use these types of digital (transaction and communication) technologies to support (rather than hinder) the digital innovation processes we are trying to promote.

? DISCUSSION QUESTIONS

The following discussion questions are relevant to using this chapter in teaching exercises and discussions or for revision:

1. What is the difference between implicit and explicit digital connectivity?

2. In what sense is digital technology not sufficient to support commercial opportunities for people in remote communities who want to engage in e-commerce?

3. What are some of the benefits and problems associated with using social media within an organisation to promote innovation?

4. Why do we need to differentiate between organisationally controlled and user-controlled social media use beyond the organisational boundary?

5. What are the opportunities and threats associated with the peer-to-peer sharing economy?

6. How do you feel your constant connectivity to other people through digital technology is affecting your life and what impact might this have on innovation?

Case questions

The following case questions might also be relevant to using this chapter in teaching exercises and discussions:

1. Thinking about the two Taobao Villages, what lessons do you draw for other remote communities that are attempting to use e-commerce to provide new economic opportunities?

2. Which of the two villages do you think is likely to see more long-term improvement for locals and why?

3. The two villages relied heavily on local leadership to open opportunities to others. What implications do you think this has for national and regional policy-makers?

4. While the two cases are presented as success stories, can you see any potential downsides to the local communities that are described, as they become more and more reliant on online retailing?

 Additional suggested readings

Boudreau, K. J., and Lakhani, K. R. 2013. "Using the Crowd as an Innovation Partner," Harvard Business Review (91:4), pp. 60–69.

Choudrie, J., and Zamani, E. D. 2016. "Understanding Individual User Resistance and Workarounds of Enterprise Social Networks: The Case of Service Ltd," Journal of Information Technology (31:2), pp. 130–151.

References

Abraham, R. 2006. "Mobile Phones and Economic Development: Evidence from the Fishing Industry in India," *Information and Communication Technologies and Development, 2006. ICTD'06.*: IEEE, pp. 48–56.

Alavi, M., and Leidner, D. E. 2001. "Review: Knowledge Management and Knowledge Management Systems: Conceptual Foundations and Research Issues," *MIS Quarterly*), pp. 107–136.

Baptista, J., Wilson, A. D., Galliers, R. D., and Bynghall, S. 2017. "Social Media and the Emergence of Reflexiveness as a New Capability for Open Strategy," *Long Range Planning* (50:3), pp. 322–336.

Bayus, B. L. 2013. "Crowdsourcing New Product Ideas over Time: An Analysis of the Dell Ideastorm Community," *Management Science* (59:1), pp. 226–244.

Boudreau, K. J., and Lakhani, K. R. 2013. "Using the Crowd as an Innovation Partner," *Harvard Business Review* (91:4), pp. 60–69.

Champoux, V., Durgee, J., and McGlynn, L. 2012. "Corporate Facebook Pages: When "Fans" Attack," *Journal of Business Strategy* (33:2), pp. 22–30.

Chesbrough, H. W., and Garman, A. R. 2009. "How Open Innovation Can Help You Cope in Lean Times," *Harvard Business Review* (87:12), pp. 68–76.

Choudrie, J., and Zamani, E. D. 2016. "Understanding Individual User Resistance and Workarounds of Enterprise Social Networks: The Case of Service Ltd," *Journal of Information Technology* (31:2), pp. 130–151.

Cohen, W. M., and Levinthal, D. A. 1990. "Absorptive Capacity: A New Perspective on Learning and Innovation," *Administrative Science Quarterly* (35:1), pp. 128–152.

Cui, M., Pan, S. L., Newell, S., and Cui, L. 2017. "Strategy, Resource Orchestration and E-Commerce Enabled Social Innovation in Rural China," *The Journal of Strategic Information Systems* (26:1), pp. 3–21.

Davison, R. M., Ou, C. X., and Martinsons, M. G. 2013. "Information Technology to Support Informal Knowledge Sharing," *Information Systems Journal* (23:1), pp. 89–109.

Deng, X., Joshi, K., and Galliers, R. D. 2016. "The Duality of Empowerment and Marginalization in Microtask Crowdsourcing: Giving Voice to the Less Powerful through Value Sensitive Design," *MIS Quarterly* (40:2), pp. 279–302.

Digital Brain Switch (2015). The Digital Brain Switch InfoGraphic Brochure. Available at: http://www.scc.lancs.ac.uk/research/projects/DBS/dbs-exhibit/dbs.html.

Durcikova, A., Fadel, K. J., Butler, B. S., and Galletta, D. F. 2011. "Research Note-Knowledge Exploration and Exploitation: The Impacts of Psychological Climate and Knowledge Management System Access," *Information Systems Research* (22:4), pp. 855–866.

Edelman, B. G., and Luca, M. 2014. "Digital Discrimination: The Case of Airbnb. Com. Working Paper 14-054." Cambridge, MA: Harvard Business School.

Faraj, S., Jarvenpaa, S. L., and Majchrzak, A. 2011. "Knowledge Collaboration in Online Communities," *Organization Science* (22:5), pp. 1224–1239.

Gani, A. 2015. *Amazon sues 1,000 'fake reviewers'*. The Guardian, October 18. Available at: https://www.theguardian.com/technology/2015/oct/18/amazon-sues-1000-fake-reviewers.

Goh, K.-Y., Heng, C.-S., and Lin, Z. 2013. "Social Media Brand Community and Consumer Behavior: Quantifying the Relative Impact of User-and Marketer-Generated Content," *Information Systems Research* (24:1), pp. 88–107.

Greer, C. R., and Lei, D. 2012. "Collaborative Innovation with Customers: A Review of the Literature and Suggestions for Future Research," *International Journal of Management Reviews* (14:1), pp. 63–84.

Gulati, G., Huang, J., Marabelli, M., Weiss, J., and Yates, D. 2015. "Determinants of Mobile Broadband Affordability: A Cross-National Comparison," *European Conference on Information Systems (ECIS)*.

Hansen, M. T., Nohria, N., and Tierney, T. 1999. "What's Your Strategy for Managing Knowledge?," in *The Knowledge Management Yearbook 2000–2001*, J.A. Woods and J. Cortada (eds.). Boston, Oxford, Auckland, Johannesburg, Melbourne, New Delhi: Btterworth Heinemann, pp. 55–69.

Henry, R. (2015). How to fake a bestseller. Sunday Times, October 11. Available at: https://www.thetimes.co.uk/article/how-to-fake-a-bestseller-chql7p52ztj.

Hildebrand, C., Häubl, G., Herrmann, A., and Landwehr, J. R. 2013. "When Social Media Can Be Bad for You: Community Feedback Stifles Consumer Creativity and Reduces Satisfaction with Self-Designed Products," *Information Systems Research* (24:1), pp. 14–29.

Hu, N., Liu, L., and Sambamurthy, V. 2011. "Fraud Detection in Online Consumer Reviews," *Decision Support Systems* (50:3), pp. 614–626.

Huang, J., Baptista, J., and Galliers, R. D. 2013. "Reconceptualizing Rhetorical Practices in Organizations: The Impact of Social Media on Internal Communications," *Information & Management* (50:2), pp. 112–124.

Kaplan, A. M., and Haenlein, M. 2010. "Users of the World, Unite! The Challenges and Opportunities of Social Media," *Business horizons* (53:1), pp. 59–68.

Kim, H.-W., Chan, H. C., and Gupta, S. 2007. "Value-Based Adoption of Mobile Internet: An Empirical Investigation," *Decision Support Systems* (43:1), pp. 111–126.

Kogut, B., and Zander, U. 1996. "What Firms Do? Coordination, Identity, and Learning," *Organization Science* (7:5), pp. 502–518.

Kuegler, M., Smolnik, S., and Kane, G. 2015. "What's in IT for Employees? Understanding the Relationship between Use and Performance in Enterprise Social Software," *The Journal of Strategic Information Systems* (24:2), pp. 90–112.

Leonard-Barton, D. (1995). *Wellsprings of knowledge; Building and sustaining the sources of innovation*. Harvard Business School Press, Boston MA.

Leonardi, P. M., Huysman, M., and Steinfield, C. 2013. "Enterprise Social Media: Definition, History, and Prospects for the Study of Social Technologies in Organizations," *Journal of Computer-Mediated Communication* (19:1), pp. 1–19.

Leong, C. M. L., Pan, S.-L., Newell, S., and Cui, L. 2016. "The Emergence of Self-Organizing E-Commerce Ecosystems in Remote Villages of China: A Tale of Digital Empowerment for Rural Development," *MIS Quarterly* (40:2), pp. 475–484.

Majchrzak, A., Faraj, S., Kane, G. C., and Azad, B. 2013. "The Contradictory Influence of Social Media Affordances on Online Communal Knowledge Sharing," *Journal of Computer-Mediated Communication* (19:1), pp. 38–55.

Majchrzak, A., and Malhotra, A. 2013. "Towards an Information Systems Perspective and Research Agenda on Crowdsourcing for Innovation," *The Journal of Strategic Information Systems* (22:4), pp. 257–268.

Marabelli, M., and Markus, M. L. 2017. "Researching Big Data Research: Ethical Implications for Is Scholars," *Americas Conference on Information Systems (AMCIS)*, Boston, MA.

McAfee, A. P. 2006. "Enterprise 2.0: The Dawn of Emergent Collaboration," *MIT Sloan Management Review* (47:3), p. 21.

Nahapiet, J., and Ghoshal, S. 1998. "Social Capital, Intellectual Capital, and the Organizational Advantage," *Academy of Management Review* (23:2), pp. 242–266.

Newell, S., and Marabelli, M. 2014. "Knowledge Management," in *Information Systems and Information Technology, Computer Science Handbook*, A.B. Tucker (ed.). Chapman and Hall, pp. 17.11–17.21.

Newell, S., and Marabelli, M.2014. "The Crowd and Sensors Era: Opportunities and Challenges for Individuals, Organizations, Society, and Researchers," *International Conference on Information Systems (ICIS)*, Auckland, NZ.

Orlikowski, W. J., and Scott, S. V. 2014. "What Happens When Evaluation Goes Online? Exploring Apparatuses of Valuation in the Travel Sector," **Organization Science** (25:3), pp. 868–891.

Satyanarayanan, M. 2001. "Pervasive Computing: Vision and Challenges," *IEEE Personal Communications* (8:4), pp. 10–17.

Schlagwein, D., and Bjorn-Andersen, N. 2014. "Organizational Learning with Crowdsourcing: The Revelatory Case of LEGO," *Journal of the Association for Information Systems* (15:11), pp. 754–778.

Schor, J. *Debating the Sharing Economy*. Great Transition Initiative, Available at: https://www.greattransition.org/publication/debating-the-sharing-economy.

Skågeby, J. 2010. "Gift-Giving as a Conceptual Framework: Framing Social Behavior in Online Networks," *Journal of Information Technology* (25:2), pp. 170–177.

Skaletsky, M., Galliers, R. D., Haughton, D., and Soremekun, O. 2016. "Exploring the Predictors of the International Digital Divide," *Journal of Global Information Technology Management* (19:1), pp. 44–67.

Sørensen, C., and Landau, J. S. 2015. "Academic Agility in Digital Innovation Research: The Case of Mobile ICT Publications within Information Systems 2000–2014," *The Journal of Strategic Information Systems* (24:3), pp. 158–170.

Stein, M.-K., Jensen, T. B., and Hekkala, R. 2015. "Comfortably 'Betwixt and Between'?," *International Conference on Information Systems (ICIS)*, Forth Worth, TX.

Treem, J. W., and Leonardi, P. M. 2013. "Social Media Use in Organizations: Exploring the Affordances of Visibility, Editability, Persistence, and Association," *Annals of the International Communication Association* (36:1), pp. 143–189.

Trkman, P., and Desouza, K. C. 2012. "Knowledge Risks in Organizational Networks: An Exploratory Framework," *The Journal of Strategic Information Systems* (21:1), pp. 1–17.

von Hippel, E. 1978. "Successful Industrial Products from Customer Ideas," *Journal of Marketing* (42:1), pp. 39–49.

Von Krogh, G. 2012. "How Does Social Software Change Knowledge Management? Toward a Strategic Research Agenda," *The Journal of Strategic Information Systems* (21:2), pp. 154–164.

Whelan, E. 2007. "Exploring Knowledge Exchange in Electronic Networks of Practice," *Journal of Information Technology* (22:1), pp. 5–12.

Yuan, Y. C., Zhao, X., Liao, Q., and Chi, C. 2013. "The Use of Different Information and Communication Technologies to Support Knowledge Sharing in Organizations: From E-Mail to Micro-Blogging," *Journal of the American Society for Information Science and Technology* (64:8), pp. 1659–1670.

9 OPPORTUNITIES AND CHALLENGES FOR INNOVATION RELATED TO IMPLICIT DIGITAL CONNECTIVITY

Summary

This chapter considers innovative processes underpinning the use of large datasets. This includes personal information to profile individuals, analyse consumer behaviours and create predictive scenarios. The peculiarity of these innovation processes is that data are analysed without the explicit intent of the data creator (e.g., a social media user) or captured without the awareness of the individual (e.g., people 'caught on camera' on a phone or CCTV). This is why we contrast 'explicit digital connectivity' as examined in the Chapter 8 with 'implicit digital connectivity' that we discuss here. Implicit digital connectivity has incredible potential to promote innovation, one example being healthcare research based on data captured with wearable devices. However, here we also focus on potential negative consequences of using implicit digital connectivity in ways that might undermine tolerance, create unfair discriminations, lead to technology dependency and more generally erode our privacy. These potential issues highlight how and why we need to consider the ethical implications of all types of innovation – an issue that is all too often *not* considered in discussions of digital innovation processes and to which we will return in Chapter 10.

Learning Objectives

The learning objectives for this chapter are to:

1. Understand the potential of implicit digital connectivity to create innovative products and services

2. Appreciate the role of implicit digital connectivity in creating opportunities for companies to improve their marketing strategies

3. Recognise the benefits of implicit digital connectivity for consumers (e.g., involving price reduction of services or free services if allowing access to personal data)

4. Understand the difference between digital innovation that involves using 'big' data and 'little' data

5. Reflect on the 'dark side' of innovations involving implicit digital connectivity.

Case: Sensors in the automotive industry

Sensor-based technologies are now widely diffused, and most people worldwide will interact with this type of technology in some aspect of their everyday life. Sensors are digital devices with the ability to record people's whereabouts (when they 'check in' with social media apps, for instance), physical movements and vitals (pedometers and the more recent Fitbit devices that can monitor exercise, heartbeat, sleep time and the like), monitoring devices at the workplace such as RFID (radio-frequency identification) tags that employees carry so that their location is known to others and so on.

The two cases introduced here discuss the so-called 'black-box', a sensor device installed in cars by automotive insurance companies and that is aimed at collecting data about a person's driving style to assess their risk of accidents and therefore gauge the appropriate amount of annual premium (the fee that drivers need to pay to be insured). This use of sensors has recently created a new insurance business model, called PAYD (pay as you drive). Vehicles must be equipped with, or connected to, technologies that can collect such risk-related factors as vehicle speed, rapid acceleration/deceleration, impact occurrences, times of the day when the car is being driven and vehicle location. The PAYD programmes currently in place use variable assortments of onboard sensors, wireless communications and global positioning system (GPS) components. The black-box is integrated with a vehicle's onboard diagnostics (OBD) system to record and transmit data on vehicle operation. Some PAYD implementations use this sensor-based system in tandem with GPS to determine vehicle location and adherence to legal speed limits. While the specific technologies employed vary, the digitally intensive nature of the PAYD model is universal. Here, we review how two prominent automotive insurance firms in Europe (Generali Italia) and the USA (Progressive Insurance) have strategically adopted PAYD provision.

Generali Italia (EU): Generali Italia is the principal operating company of Generali Group, a leading player in insurance and financial markets around the world. Operating in more than 60 countries, Generali Group focuses primarily on the life insurance segment. However, the firm's non-life segments (including automotive, home and health insurance) have become increasingly prominent. Generali Italia (hereafter, Generali) is a publicly traded firm headquartered in the north-eastern Italian city of Trieste. The firm has a prominent position in the financial industry in Italy. In 2013, Generali Italia enjoyed the second-largest share of the Italian insurance market in both the life (15.1 per cent) and non-life (18.1 per cent) product segments. With more than 10 million policyholders and 7,200 employees, Generali relies on a salesforce of 3,200 agents and more than 21,500 sub-agents. In the automotive insurance segment, Generali has emphasised the adoption of black-box telematic technologies (telematics is a combination of telecommunications and informatics and refers to the combined use of telecommunications – for instance, cell phone wireless, with digital devices) in recent years. Generali launched this product in collaboration with its sister company, Genertel, an online insurance company owned by the Generali Group. Both consumer/market conditions and regulatory factors have driven Generali's proactive adoption of black-box innovation.

Generali's approach to PAYD centred on adopting a sensor-based OBD device that can record several significant details of individuals' driving behaviour. The device is equipped with a GPS system and a subscriber identification module (SIM) card that transmits data every day to a third-party telematic service provider (TPS) data centre. The TPS transmits a summary of the data collected to Generali on a monthly or even daily basis. As noted above, the data collected include details on acceleration/deceleration, speed travelled, vibration and impact events. Interestingly, in the case of an accident (i.e., detected as an abnormal deceleration or an impact), data are transmitted to the TPS and on to Generali immediately. Emergency services are automatically alerted and given the GPS location of the insured vehicle. In the case of a car accident that does not involve emergency services (i.e., not detected by the OBD device), Generali might ask the TPS for specific data (e.g., time of day, GPS coordinates) to address potential ambiguities surrounding the accident or the possibility of fraudulent claims. Once a policyholder opens a claim, Generali requests data such as date, location, speed and impact. If the information that the TPS provides is inconsistent with that which the policyholder provides, Generali may initiate antifraud procedures. According to top management at Generali, in 2014, one of the clearest advantages associated with their PAYD model was the ability to quickly investigate potential fraud incidents.

Generali's business model reflects a clear demarcation of roles and responsibilities for the various organisational parties involved. The TPS wholly owns the digital elements involved (i.e., OBD, GPS, SIM card and data centre). Thus, Generali has fully outsourced the telematics service. Generali then uses the data collected through the telematics service to evaluate driver behaviour, assess risk profiles and determine policyholders' premiums. If the algorithm that Generali uses to identify good and bad drivers' changes, and Generali determines that it needs additional data to appropriately assess driver behaviour, it needs to establish a revised agreement with the TPS. In short, the TPS sells data at a flat rate to Generali – but only a certain amount of data and on a pre-established basis. Figure 9.1 shows the sensor-based technology and underlying processes that support Generali's strategy.

The black-box technology offers several purported benefits for Generali's consumers. Most notably, consumers experience lower insurance premiums if the algorithmic analyses determine their driving styles to be 'good'. In addition, Generali sends a monthly report with key indicators of consumers' driving behaviour, such as occurrences of rapid deceleration (i.e., hard braking suggestive of attention concerns) or instances of driving over the speed limit. These reports provide feedback to consumers to enable them to 'learn' how to drive more safely and, accordingly, reduce their insurance premiums and increase their security. Consumers also benefit from the ancillary services related to the telematics data (outlined in the contracts) such as roadside assistance, automated notification of emergency services and assistance in carjacking incidences.

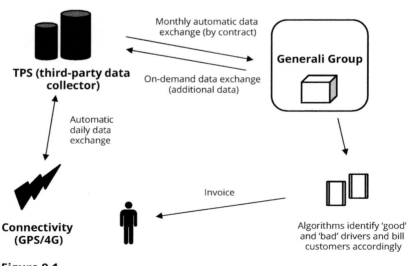

Figure 9.1

Generali's PAYD model, technology and processes

Generali has identified three primary benefits from the telematics technology. First, most good drivers choose to have an OBD because they know that they will save money, which creates a self-selection mechanism for lower-risk drivers. If they can retain these responsible consumers, early adopters such as Generali will in all likelihood have better drivers than their competitors, which will enhance their profits. Second, the black-box technology enables Generali to sell more insurance policies because of the ancillary services provided. Third, the telematics technology dramatically enhances Generali's fraud detection capabilities by expanding the evidentiary base available for analyses.

Progressive Insurance (USA): Headquartered in Mayfield Village, Ohio, Progressive Insurance is the country's fourth-largest automotive insurance provider with over 15 million active policies and over 8 per cent of the national market share. The firm offers coverage for a wide range of vehicle types, including personal and commercial cars, motorcycles, recreational vehicles and boats. Progressive sells policies and interacts with its consumers on multiple platforms. In its rise to the top ranks of the US automotive insurance marketplace, Progressive has gained a reputation for both process and technological innovation. The firm made early use of databases to analyse accident and client data in more detail than its competitors, which enabled it to profitably serve the non-standard market. In the late 1980s, Progressive invested heavily in its information infrastructure by installing a large computer system to accelerate both claims and application processing. This investment resulted in fewer inaccurately priced policies and enhanced the productivity of claims adjusters. In the 1990s, the company embarked on several innovation initiatives, including introducing comparative pricing (i.e., providing prospective consumers with quotes from Progressive and up to three competitors), developing a mobile claims processing process and launching the industry's first commercial website.

More recently, Progressive has become one of the first US-based insurance companies to adopt the PAYD model. The primary focus of Progressive's current PAYD initiative is a programme called Snapshot. Snapshot is a small sensor-equipped device and voluntary discount programme that enables drivers to reduce their premium payments by sharing driving data with Progressive. Originally piloted in 2003 (then under the name TripSense), Progressive rolled out Snapshot in earnest in 2011.

At the heart of the Snapshot programme is the Snapshot device, a small item of hardware that is plugged into the OBD port located under a vehicle's dashboard. The device records data on several key aspects of driver behaviour, including time of day driven, vehicle speed, and rapid acceleration and braking. The device transmits data back to Progressive via an embedded wireless modem, and device malfunctions are communicated to the driver in the form of email alerts. The earliest phase of Progressive's PAYD initiative incorporated a GPS component that

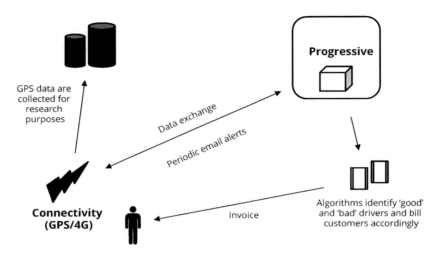

Figure 9.2

Progressive's PAYD model, technology and processes

contributed to a driver's profile. However, it eliminated the GPS element in 2011 in response to consumer privacy concerns. It later partially reintroduced the GPS element, but the company no longer uses its data to establish a consumer's driving profile and rather uses it for research-and-development purposes, explicitly stating that GPS data are not is used to calculate insurance premiums. While most of the technical characteristics of Progressive's PAYD system are similar to those of Generali, the former owns the whole IT infrastructure. Figure 9.2 highlights Progressive's sensor-based technology and underlying processes.

The Snapshot programme proposes to offer consumers greater control over the premiums they pay. Participation in the programme is entirely voluntary and consumers can opt out at any time. In terms of pricing, an individual's standard premium (i.e., the premium based on traditional underwriting inputs) acts as a baseline. By participating in the programme, policyholders can receive discounts of up to 30 per cent based on their driving data, but the average discount achieved is around 10 per cent. In addition to the financial benefits, participation in the Snapshot programme provides a consumer with greater understanding of their driving habits. After 30 days in the programme, participants can log into Progressive's consumer portal to review the data gathered on their device. The company also provides tips for how drivers could enhance their discounts through safer driving behaviours.

The Snapshot programme presents multiple benefits to Progressive. Most fundamentally, collecting actual usage data holds the promise of greatly improving how the company assesses risk. By knowing when a vehicle is driven, a vehicle's total time on the road and specific driving behaviours, Progressive can more accurately price a driver's policy relative to the underwriting risk that the driver represents. In addition, by raising the visibility of risky driving behaviours among the drivers themselves, the Snapshot programme may prompt Progressive policyholders to engage in safer driving habits. Additionally, the Snapshot programme has arguably enhanced consumer satisfaction. Progressive's customer satisfaction ratings with respect to both pricing and policy offerings are significantly higher among Snapshot participants than among those who do not participate in the programme.

The cases are used throughout this chapter to help guide a discussion of key topics and themes.

Introduction: Data and the digital world

The twenty-first century has witnessed the widespread diffusion of digitised objects that capture data about the minutiae of our everyday lives (Hedman et al., 2013). These digitised objects are everywhere, a phenomenon described as *ubiquitous computing* (Nolan, 2012), as noted in Chapter 8. This trend is associated with the so-called internet of things (objects are connected to each other and to people as well as simply people connecting to each other over the internet) but is also related to the massive growth in the use of all the different types of computing technology (often mobile) that we now use to connect with other people and organisations (Sørensen and Landau, 2015). The data trail we leave from using computing devices and other digitised objects is increasingly used by companies to target and personalise the information that we receive on products and services, based on developing algorithms that can make predictions about individuals by recognising complex patterns in huge datasets compiled from multiple sources. For instance, in the case, we have described how car insurance companies are innovating by providing customers with OBD equipped with GPS and sensor technology that captures many aspects of driver behaviour. These data are then used by the insurance companies in their decision-making, based on developing algorithms that can predict whether a person is a safe or risky driver and so more or less likely to have an accident. Personalised insurance premiums are then set based on these data-driven predictions. In other words, innovative business models are being developed around the ability to use digital technologies to monitor actual activity.

Similarly, data collected from social media platforms are used to gauge users' opinions on products and services, political views, interests, hobbies and the like. These data are then processed by algorithms that are used to prioritise certain contents that are made available to different people. In other words, algorithms 'decide' whether, for example, we are a driver at risk of accidents and whether we see, on our news feeds, a Walmart cold-cure advertisement, a request to sign a change.org petition and/or our former high schoolmate's pictures of her newborn son.

The rise of datafication

The phenomenon of datafication, where communications and actions are turned into data (Newell and Marabelli, 2015), has been stimulated by the increased storage capacity of modern computer chips. It was introduced into our everyday lives in a subtle way (companies or other organisations tended not to advertise this aspect of their operation) and most individuals found algorithmic decision-making a good way to save money (such as in the case of car insurance policies) or to have 'likable' content readily available on social media. Similarly, companies benefitted from being able to better profile their customers and, in so doing, were able to shift their marketing focus from traditional 'market research' (e.g., individual surveys or focus groups) to data mining research on social media and, more broadly, online 'digital traces' (Venturini and Latour, 2009) – the digital footprints left by internet surfers who upload personal information on various websites. Political parties likewise have made heavy use of this type of digital profiling to target their advertising in an attempt to persuade people to vote for them or not vote for another party (Bode, 2012; Tait, 2017) (Figure 9.3).

More recently, academics, the media and governments (lawmakers and more generally politicians) have highlighted potential negative societal consequences associated with the intense use of digital traces. There are two major concerns related to this 'datafication', which we define as the processing and

Figure 9.3

Expanded computer storage is one of the factors enabling datafication
©*Getty Images*

analysis of vast amounts of data (so-called big data) with the aim to distil relevant insights to support and/or automate decision-making. It involves data captured with traditional techniques (from databases, cf. business analytics) but also through sensor technologies and web scraping of social media sites. First, the processes underpinning data collected through computing devices and digitised objects are often not fully disclosed to the data creators (i.e., the individual user) or the 'walking data generators', as Brynjolfsson and McAfee (2014) describe them. This is why we associate this phenomenon with *implicit* connectivity, in contrast to *explicit* connectivity. Explicit connectivity involves crowdsourcing and other forms of innovation users are willing to engage with (as we saw in Chapter 8). Second, the data collected by companies is processed by algorithms that make decisions on the basis of data analytics processes. For instance, people's behaviours on social media such as Facebook (e.g., likes and friends) has been shown to be a good predictor of their ability to pay off loans, hence some companies are using this data to make decisions on loan offerings (Huang et al., 2017). While the link between loan repayments and social media use is statistically relevant (and as a result represents meaningful information for financial institutions), it might also lead to what might be considered unfair discrimination. For example, it might make it more difficult for some individuals to obtain loans if they are not social media users and are thus unable to prove their solvency (at least through this method). Moreover, trends are based on statistics that reveal relationships, but there will always be exceptions to these relationships. Therefore, even a social media user may be unfairly discriminated against if his or her social media use is associated with high-risk borrowers while they may not in fact be a reckless borrower themselves. While, traditionally, these exceptions have been managed by human beings (e.g., a bank manager could offer a loan even if a credit rating score is low, based on their assessment), if organisations rely on algorithms, it is a computer that decides whether or not a loan can be accessed. This is a major ethical concern and we will address it along with other concerns in the remainder of the chapter and in the last chapter. We do so by illustrating that innovation associated with data-driven decision-making poses social and ethical issues because there can be a difference between how business is benefiting and how individuals or society are benefiting – or otherwise. We first identify two key sources of data (computing devices and digitised objects) and, second, provide examples and discuss a variety of consequences.

Overall, implicit digital connectivity (and associated data analytics and algorithmic decision-making) is an innovative technique to capture and examine vast data sources and develop new opportunities from this. We ultimately conclude that it is important to understand how to use cutting-edge digital innovations in responsible (rather than reckless) ways, therefore exploiting their business and societal values while constantly monitoring ethical issues. Of course, ethical considerations are important for all types of innovation, so the issues we raise here are much more broadly applicable to any analysis of innovation processes. We return to this issue in Chapter 10. We now turn to consider the unique aspects of current innovations surrounding digital technologies.

Algorithmic decision-making and big and little data

In Chapter 8, we saw how more extensive explicit digital connectivity has allowed new ways to innovate by, for example, allowing companies to tap into the 'wisdom of the crowd' (Martinez and Walton, 2014) and increase involvement in decision-making within an organisation. More importantly, aside from the data explicitly contributed by users in this era of ubiquitous computing, there is also the data trail left by our computing application use (e.g., locational information and content) as well as from other devices that increasingly have tracking and sensing software in them so that 'the digital artefacts will be able to remember where they were, who used them, the outcomes of interactions, etc.' (Yoo, 2010, p. 226). We must comprehend that each of us is now a walking 'data generator'. This data trail then provides more immediate access to what we are doing and how we are using products and services that can result in tailored or personalised advertising and the development of new products or services. In the past, companies might have conducted a focus group to collect data about customer reactions to a new product. Now, if the product embeds tracking sensors, a company can actually gain access to real-time data on how their product is being used. For example, with e-books, a publishing company can instantly track what type of customers are downloading a new magazine title and what pages they are dwelling on, rather than having to wait for bookshops or customer survey instruments to provide this information. This can allow the company to quickly decide how it might change the magazine format if certain parts are not being read, or it may decide on a different marketing campaign if it realises that the readership is different from its original idea of what market segment this magazine would attract (Figure 9.4).

It is this data trail that provides the opportunity for organisations to move to data-driven or algorithmic decision-making. Algorithmic decision-making is based on collecting large quantities of data from the tracking software that is now built into the applications and devices that we use in our daily lives and then developing algorithms that make a selection in order to model a particular phenomenon of interest. Examples include the previously mentioned Facebook likes and friends being used to predict a person's credit risk or OBD data being used to determine a driver's risk of an accident. Thereby, data can be used to track general trends (big data) as well as the minutiae of an individual's everyday life (little data).

Little data (also known as smart data) refer to all the data that is today collected at the individual level. Data is collected through digital traces from any kind of computing device that a person uses or from the objects with inbuilt

Figure 9.4

Algorithms enable decision-making capabilities
©*Getty Images/iStockphoto*

sensors that individuals carry or use and that record (and in some cases transmit to data centres) information related to their whereabouts (locator-based or GPS-equipped sensors), their activities (e.g., use of an e-book) and even more sensitive information such as health data (heartbeat, blood pressure etc. from the health monitors that are increasingly being worn). An example of little data relates to data collected through a car's black-box that, as noted in the cases with which we opened the chapter, is now offered by insurance companies.

Datafication and comparing big and little data

Big data refer to all the data collected from us as individuals that can then be combined to allow organisations to see connections in those data that might suggest ideas for new products or services or, more importantly, how to target products and services to particular individuals. The characteristics of big data have been described as 3Vs – volume, variety and velocity – with the crucial point being that the data are collected from millions of individuals and from many different sources and can be brought together and processed at high speed, therefore allowing complex statistics to be used to explore connections in almost real time in order to make predictions. Developing algorithms that analyse big datasets can then identify behaviours and characteristics that predict certain outcomes. As we have seen with OBD, data are collected at the individual level (the individual's car-driving behaviour is tracked) and then combined with other individual driver behaviour data, plus potentially data from other sources (such as accident reports), and an algorithm is written that predicts behaviours associated with 'good' and 'bad' driving (i.e., with behaviours that are associated with more or fewer accidents). Individuals can then be compared against this algorithmic standard and deemed to be good or bad drivers – with a consequent upward or downward adjustment to their insurance premium.

Decision-making based on little data is more sophisticated than traditional, general trend, approaches to using data, and we shall demonstrate this by returning to the car insurance example that was the focus of the cases with which we opened the chapter. Car insurance companies have long been able to use police statistics (general trend data) to discriminate in terms of the premiums paid (Lemaire, 2012). For instance, in the past, the industry has charged men higher premiums because the data on general trends indicate that they drive less safely, on average, than women. Such data-driven decision-making has been questioned because it can go against the ethical principle of equal or fair treatment. This is exemplified in a 2012 EU legal decision which meant that insurers can no longer use statistical evidence about general gender differences to set premiums. Therefore, although gender differences *are* clear from the data (e.g., young male drivers are ten times more likely to be killed or injured on the road than those – of both sexes – over the age of 35), it is considered to be discriminatory to use this evidence to differentiate premiums for men and women. The point about this change in the law is that it was considered to be discriminatory because while young men *in general* may drive more recklessly and so be more prone to accidents, an *individual* young man may not and would therefore be discriminated against when insurers set premiums based on group trends observable from general trend data.

The idea of using little data is that it moves from the general to the specific – the actual individual and their behaviour. So, by using data from the OBD in a specific car, the insurer would not be setting premiums based on general trends in accident rates between groups but instead would base their calculations on the actual driving habits of an individual. In this way, people are held accountable for their own (individual) behaviours, which is fairer than simply categorising them based on their group identification – in this case, whether male or female.

However, there remain concerns about this use of little data. For instance, in relation to the OBD example and its use in the automotive insurance industry to innovate in relation to setting premiums, individuals might not want to be monitored and thus would not want to sign up for having OBD installed in their cars. However, for those financially less well off, they might feel forced to give up their privacy (and as a result 'accept' being monitored) because they might not be able to afford car insurance otherwise. Acquisti

Figure 9.5

Onboard computers are able to capture the minutiae of our life when we drive our car

©Getty Images/iStockphoto

(2012) describes this as the monetisation of privacy. We will return to this issue of privacy later in this chapter (Figure 9.5).

McAfee et al. (2012) argue that this type of algorithmic decision-making is superior to traditional 'HiPPO' (highest-paid person's opinion) decision-making, essentially because human decision-makers have biases. McAfee and colleagues argue that algorithmic decision-making – here defined as an algorithm's 'power' to lead (instead of simply influence) decisions or even to act autonomously and make automatic decisions on behalf of human beings – is often *superior* to human judgment-based decisions given the latter's inherent biases. However, not only does this ignore the fact that it is humans (with their biases) that create the algorithms in the first place, in this chapter we also more generally consider some of the negative (as well as positive) consequences associated with innovations related to the datafication of our world and associated algorithmic decision-making. We consider this first in relation to the implicit data collected through our computing application use and then the implicit data collected by sensors in everyday objects.

Key Concepts: Big and Little Data

Big data:
Using a variety of sources of data to explore connections that can identify trends able to provide myriad opportunities for many different types of new products and services as well as research. While individuals are the source of the data, their specific information is removed from the analyses in a bid to identify general trends – e.g., what activity characteristics, as measured by all those using wearable health monitors, are associated with reduced chance of a heart attack, as measured by health record data

Little data:
Also known as smart data. Based on big data analysis, individual data can be used to identify specific services or products that might be attractive to a particular individual. Data used in this way compare the individual's data activity with general trends to target and personalise – for example, based on knowing from big data analysis what activity characteristics reduce the chance of heart attack, individuals who use a wearable health monitor can be sent alerts to maintain activity levels associated with such a reduced chance of heart attack

Implicit digital connectivity and computing applications

As we have implied, the widespread use of computing applications allows various types of algorithmic analyses that can suggest trends and make predictions. For instance, Facebook likes' are able to shed light on individuals' personality traits, political views, sexual orientation and other behavioural characteristics (Ross et al., 2009; Youyou et al., 2015) as well as their likely propensity to repay a loan. These data are used in aggregate to, for instance, suggest that people, over the summer, are less stressed or that a geographical region is dominated by liberal (or conservative) ideas. Big data again might be valuable here to identify general trends. For instance, in November 2016, conversation analysis of over 4 million Facebook news feeds led to predictive results associated with the US presidential election that were more accurate than traditional polling techniques. This opens interesting avenues for innovative ways to understand collective phenomena, such as a general election or changing social opinions. However, we must consider that the Facebook evidence emerged as a 'post-hoc' analysis of the 2016 US presidential elections, so this was not a predictive analysis. Instead, predictive analyses relying on big data had suggested that Hillary Clinton would win, which of course turned out to be incorrect and thus questioned the reliability of big data, at least in this particular context and for predictions rather than post-hoc analysis. But the 'power' that rests on social media data (and their interpretation) goes beyond polling results and can be used to map and predict phenomena that relate to the safety and health of entire countries, for example.

However, the predictive power of computing applications and big data is far from perfect. For instance, in 2013, Google engaged in a worldwide programme to try to predict the spread of the winter flu in various countries by combining people's activity on the Google search engine (keywords) with the origin of the various searches (through IP addresses). The project, called GFT (Google Flu Trend), compared search findings with a historical baseline level of influenza activity for its corresponding region and then reported the activity level as minimal, moderate, high or intense (Figure 9.6).

This use of big data to predict the spread of flu represents a noble and innovative effort to understand how flu develops and therefore how to prevent major outbreaks – or to intervene rapidly to limit an outbreak. However, as several reports show (for example, see the 2013 article by Adam Kucharski entitled 'Google's flu fail shows the problem with big data'), Google predictions were far from being 100 per cent accurate. Among the problems were 1) the limited ability of individuals to self-diagnose flu (symptoms) and 2) the dynamic nature of the Google search algorithms which, among other things, continuously change the priority of certain results (webpages) on the basis of the terms searched and the number of visits that websites have.

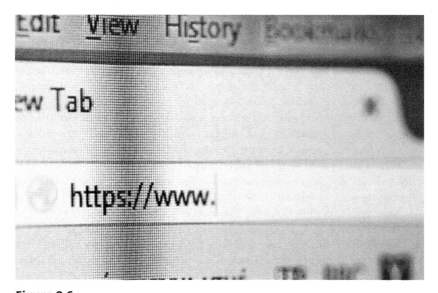

Figure 9.6

Search engines can drive innovative discoveries
©Getty Images/iStockphoto

The potential errors of big data predictions raise concerns because companies and government institutions often make decisions based on these analyses. Yet trends and statistics ignore exceptions, minorities and the like – and can be simply *wrong*. For instance, one could argue that the financial crisis of the early 2000s was at least partially a product of the inaccuracy of big data predictions, with the algorithms that were used to price mortgage-backed securities based on their predicted likelihood of default clearly not taking into account all the risks. At the same time, these algorithms were not questioned because the basis of the algorithms was neither clear nor easily accessible, either to the senior managers in the financial institutions where the algorithms were being used or to the credit rating agencies that were evaluating these products (Clark and Newell, 2013). This reminds us that, while big data are 'simply' a very large amount of data collected from various sources, the algorithms that manipulate the data are extremely complicated and often even their creators struggle in understanding why any particular algorithm was able to come up with a specific prediction.

Ambiguity and algorithms

The ambiguities affecting how algorithms operate have two root causes. First, algorithms are set up to compare and combine very different types of data, and look at historical data trends, which might be invisible to human beings (because of the vast amount of data that algorithms are able to process). For instance, the trend analysis used to predict the likelihood of a defaulting loan before the 2008 financial crisis did not take account of the fact that loan companies were reducing the formal verification of income that in the past was the basis for making loan decisions. This meant that many more people defaulted than had been predicted, with the number of 'bad debts' on the books of financial institutions soaring (Clark and Newell, 2013). Second, algorithms 'learn', so what an algorithm can do at the time of its creation can vary from what the same algorithm will do after being 'trained' with real data. One key problem with this learning ability of algorithms is that they are not being trained to be 'ethical'. In other words, when algorithms are created – by humans of course – their logic of action is typically to maximise profits, with societal consequences not typically factored into the algorithm's self-adjustment.

Not only are companies using this type of big data analysis, they are also undertaking little data analysis. However, little data can be more invasive and raise more privacy issues. With big data, (supposedly) the various data collected are divorced from their owners – a process known as de-identification (Hay et al., 2008) – therefore the data generators (e.g., social media/Google users) are warranted anonymity. Little data, instead, relate data directly with their owners. For instance, computing application sites can profile an individual's internet activity through tracking cookies. Tracking cookies are text files stored on our computers that document our internet behaviours (and usernames, passwords) enabling quick access to our email, bank accounts and the like. Tracking cookies represent the technology that lets Facebook 'know' that we have just googled flu drugs because (it is assumed) we feel sick. In turn, Facebook can customise ads on our profile accordingly. Tracking cookies therefore allow companies to provide users with personalised content.

One problem, typical of implicit digital connectivity, is that the ways algorithms decide what ads we should see on the basis of our previous internet activities are unknown to us – and often also to the data companies that use tracking cookies for online sales purposes, which is why some organisations end up advertising on websites that are not ones with which they would want to be associated. For instance, we may see many posts about the newest iPhone and assume that many of our Facebook friends are posting articles about it. However, the frequency with which we see these posts may be partially due to our having clicked an advert related, for example, to a specific smartphone. Facebook's algorithm decides that we are interested in this technology and then shows us others' posts that are related to the newest iPhone, for example. A consequence of such use of algorithms by corporations to decide *for the consumer* the posts, news or advertising that they are exposed to is that it may lead to a slow and often subtle manipulation of consumers' worldviews as well as to new forms of discrimination. Simply, what is presented to the reader is decided by an algorithm – tapping into prior searches – and is not based on an explicit personal choice, a phenomenon that can be described as *uninformed control* (Newell and Marabelli, 2015).

An example of uninformed control by a corporation that produces worrisome ethical issues is found in the account presented by Eli Pariser, the author of the book entitled *The Filter Bubble: How the New Personalized Web Is Changing What We Read and How We Think*. Pariser discusses how 'Facebook was looking at which links I clicked on, and it was noticing that I was clicking more on my liberal friends' links than on my conservative friends' links. And without consulting me about it, it had edited them [conservative friends' links] out. They disappeared' (Pariser, 2011). He describes this as a 'filter bubble', the idea that the computing application providers (e.g., Facebook or Google) are making decisions about what we like/dislike and then filtering out what they think we don't like – based on predictive algorithms of our past actions compared to the actions of millions of others – so that we are exposed only to a bubble of information that fits with our likes.

In the longer term, this manipulation by corporations of what the consuming public is exposed to – exposing us only to things that we like (or rather the things that an algorithm *assumes* we like) – may produce societal changes. For instance, our exposure to online diversity will be reduced, as in the above example of removing links to a group with a particular political orientation. A *New York Times* article (Somaiya, 2014) reported that Greg Marra, a Facebook engineer, said that, 'We think that of all the stuff you've connected yourself to, this is the stuff you'd be most interested in reading', explaining further that an algorithm monitors 'thousands and thousands' of metrics to decide what we should see on our Facebook page. These metrics include what device we use, how many comments or likes a story has received and how long readers spend on each article/post. The assumed goal is 'to identify what users most enjoy'. However, this *New York Times* article also pointed out that this practice of showing us only things that 'fit' with our little data profile limits our choices and might inhibit our capacity to make informed decisions – for example, on what we buy and even what we think.

These data analysis practices, then, are leading to citizens being exposed to less and less diversity online. A potential consequence is that we may become less tolerant of diversity in our lives, meaning that we may become less able to listen to someone who apparently may think differently to us. Moreover, there may be other, more dangerous consequences in the long term that are associated with diversity intolerance – whether related to ethnicity or political or religious belief, for example – and the increased exploitation of the vulnerable. For instance, if algorithms calculate who is less capable of making good financial decisions, personalised adverts can then be sent persuading these people to take out risky loans or high-rate instant credit options, thereby exploiting their vulnerability. The strategic use of data by corporations to personalise our internet, in other words, is a way of allowing discrimination in marketing. Organisations have always discriminated in terms of their marketing, but this new type of discrimination is pernicious because the only person who has access to the outcomes of the discrimination is the individual being discriminated against – and who is often unaware of the fact that they are exposed to discriminatory information (uninformed control, in other words), making it easy for businesses to use personalisation in a way that can harm the vulnerable.

Summary: positives and negatives of big and little data

Big and little data analytics can be successfully used to process user-generated content in digital settings, leading to innovative approaches to marketing. Specifically, this allows the targeting of more personalised offerings. This undoubtedly has some benefits, making it easier to find information, entertainment, goods and services that we might like/want, so reducing the need for us to personally filter out to find what we are looking for. On the other hand, as we have seen, there are also some downsides in that our exposure to things on the internet is increasingly filtered for us, based on algorithmic decision-making, including – potentially – making us less tolerant of ideas that are different from our own and allowing companies to target the vulnerable. We next turn to consider how innovations associated with the digitisation of objects have some potential positive but also negative consequences for individuals and for society more generally, especially in relation to our ability to learn and be independent.

Implicit digital connectivity and digitised objects

As we have seen, many objects now have inbuilt sensors that connect to the internet and so collect data that can be used in many ways. In other words, these objects are able to track what people are doing and feed this information back to some central database. For example, sensor technology can be used to help protect citizens when prisoners are released from jail but are required to wear a tracking ankle-bracelet. These systems are aimed at using new technologies in innovative ways to improve the overall security of our society, with the sensor acting as a deterrent for prisoners to escape or commit a crime when they are on parole. Other instances where security is enhanced by objects that embed sensors is the ability to trace a stolen device, or a kidnapped child, as in a case that occurred in September 2013 in Texas, where the Houston police were able to trace the whereabouts of a kidnapper by tracking the iPad he had with him in his car. A similar example relates to police authorities being able to detect a crime because it is all 'caught on tape', for example with sensor-activated security cameras (Figure 9.7).

The above examples of companies and government agencies using sensor technology to protect citizens come at some cost in terms of individual privacy, however. In terms of locating a lost smartphone, it has to be the user who, deliberately, accepts giving up her/his (right of) privacy by activating the 'find my phone' option. However, in some circumstances, one's use of social software applications affects others' privacy, as, for example, for people who are tagged in somebody's Facebook profile without them knowing. Perhaps not surprisingly, privacy advocates have argued that in these types of exchanges consumers are justified in expecting that the information they share remains private among those to whom it was originally disclosed, dependent on users' risk perceptions, rather than being shared with third parties who may subsequently behave opportunistically (Gerlach et al., 2015).

The exponential diffusion of tracking software embedded in social networks such as Facebook and the sensors in many other digital devices leads us to think that it will be hard for organisations or governments to regulate how individuals use technologies that enable tracking responsibly (e.g., in a way that balances security and privacy). For instance, several states in the USA have agreed that police officers should wear cameras after recent cases involving police officers' improper use of force. The idea is that sensor technologies are used to record interactions but that the recordings would not actually generate big data since the camera records would be reviewed only in particular circumstances (e.g., when a citizen claims that have been badly treated by a police officer). However, this and other types of sensors are pervasive and invasive, and the data (e.g., the camera records) can be stored. In such circumstances, we do not know whether in the future somebody will develop an algorithmic-based decision system to analyse the data (e.g., to

Figure 9.7

Surveillance cameras contribute to address crimes, yet they pose societal issues

©*Macmillan Education*

assess the performance of police officers or to record the behaviour of individual citizens over time).

Another innovative example of using tracking systems involves organisations that track the activities of their employees (e.g., requiring that they carry a card that tracks where they are in a building or their activity on the computer they use to undertake their work). The idea here is that such tracking can help to assess employees' productivity. More specifically, algorithms can be used to identify what activities are characteristic of more and less productive employees – algorithms associate the average productivity of individuals with, for example, their 'physical' whereabouts in the organisation, how often they take breaks and how much time they spend on different

Figure 9.8

Monitoring workers' whereabouts might increase short-term efficiency while over-stressing employees

©Getty Images/Hero Images

activities. Once the 'productive behaviours' are known, employees can be trained to work in the more productive way – or so the argument goes (cf. our earlier discussion of Frederick Taylor's Scientific Management in Chapter 1). For instance, at a large global company that we are currently studying, located on the US East Coast, data are collected that show the internet activity of each employee during breaks as well as work time. This type of data is captured through network monitors, with management then assessing performance and administering merits, or punishments, on the basis of their analysis. However, as with other types of such algorithmic decision-making, the limitation of this approach is that the algorithm does not account for 'exceptions'. For instance, a very creative employee might work intensively for a few hours but then take a longer break than the norm. While taking longer breaks might be a sign of less productive behaviour for most, for others it might lead to greater productivity (Figure 9.8).

Data, algorithms and the impact on innovation

Careful reflection on such heavy control strategies, then, raises the question of whether data help or hamper innovation. Moreover, if algorithms are used to assert that people with specific characteristics are not fit for certain jobs, algorithmic decision-making can veto hiring potentially creative talents

243

because they are 'judged' as being unreliable by an algorithm. These innovative ways to recruit may also not be seen as ethical since recruiting based on algorithms risks penalising people who can be equally good performers, but because they fall into a category that is generally seen as being less productive, they pay the consequence of algorithmic generalisation. While it is true that recruiters have always been known to be biased (e.g., recruiting people who are similar to themselves), at least humans have the ability to recognise their own biases and potentially overcome them. If decisions are left to algorithms, while they may not be biased in the same sense as human beings, they are nonetheless biased by general trends and may therefore miss opportunities to recruit those who may not conform but who are potentially more creative and innovative. In Figure 9.9, we depict a model that some companies adopt to 'follow' their employees and as a result analyse their behaviours in the workplace.

While these tracking systems may discriminate unfairly against those who work in a different yet productive way, there are other problems with such technologies in relation to dependence and learning. For instance, objects with tracking sensors allow algorithms to 'assist' people in their everyday activities. Algorithm-based GPS systems that suggest the best route to our desired destination might save us time by identifying the fastest route with respect to current traffic. However, in the long term, they might inhibit our ability to make decisions on our own, should the GPS stop working, for example. Or again, using the car example, our cars can now brake for us in the face of a hazard, potentially restricting our ability to know when and how hard we should be braking in different situations – which would clearly be

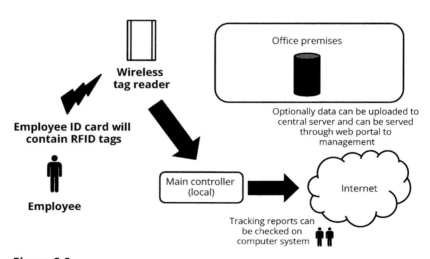

Figure 9.9

Employee tracking systems

problematic if the technology were to go wrong. These systems can thus become a crutch on which we become overly dependent, controlling our actions and in so doing making us less independent and able.

For instance, think of tracking software that can now be used in an educational context. Students can be tracked, so that whether they are attending lectures and seminars is automatically recorded, for example, when having to use a personal ID card to enter a classroom. Their use of the Virtual Learning Environment (VLE), where study materials are located that are required reading, can be monitored in terms of how frequently they have downloaded materials or how much time they have spent on the site. Additionally, the books that they take out from the library can be monitored; the book purchases they make at the college bookshop can be recorded and so on. All this information can then be gathered together to identify those students whom an algorithm identifies as not 'engaging' with the learning environment – because they are not in the classes they should be and they are not accessing the VLE or reading the required texts. They can then be asked to meet with their academic advisor with a view to modifying their behaviour or with the threat of punishment if they do not change their engagement activities. This might help nudge some students into more productive learning habits. However, it does not account for the non-typical student who, for example, might cruise through the term but who works extremely hard in the exam period and does very well compared to his or her peers who have engaged throughout the term. Just such a student might be excluded based on algorithmic decision-making when in fact their approach to learning works for them. It might also mean that students engage in the required activities not because they feel this is important for their learning but because they feel coerced into doing this. This may not bode well for their future as independent learners who must self-motivate to study and learn new things in their future careers.

We can take this one step further in terms of thinking about the potential negative societal consequences of the growth of innovations around this type of automated technology based on sensors and connections to the internet – the ability of digital technologies not simply to aid people doing a job but 'intelligent systems' that can now take over and 'do the job'. Indeed, the types of job that are at threat are not simply mechanistic jobs that robots have long performed but jobs based on cognitive skills (e.g., medical scans can now be read by a machine rather than a doctor, with the machine having been 'taught' to distinguish scans which indicate a problem from those which do not). Machine learning (or artificial intelligence) like this is now being experimented with for a whole host of different 'professional' activities. While we cannot predict the future, it is likely that such intelligent systems will, at least initially, support people in their jobs rather than completely take over, but the scope for innovation is clearly considerable, as is the threat to certain job types. Overall, then, innovations associated with data collected through digitised objects, as with those collected through computing applications, might have relevant and also worrying consequences, at the individual as well as the societal level, even while they open up interesting opportunities for making money from a purely business perspective.

☁ **Key Concepts:**

Some potential 'downsides' of data-driven decision-making digital applications and the filter bubble:

- Use of past searches, likes and friendship networks analysed to provide information and content tailored to a specific individual, with the potential that individuals will become less tolerant of people and information that are different from them or the way they think; this personalisation also opens up opportunities for exploiting the vulnerable

Digitsed objects and dependency:

- As more of the objects we use in everyday life are digitised to support/control our activities, we can become overly dependent on the technology and find it difficult to do things without the crutch of technology; our ability to learn may also be impeded; and jobs themselves may be under threat as digital technologies take over many activities

---

Innovation opportunities from big and little data

We have demonstrated in this chapter that data can be extremely valuable to companies and other organisations since, as a consequence, they can innovate using data, either in terms of how they market and sell or in terms of the actual products and services they produce. Some companies (such as Google and Facebook) have recognised the value of data and have built massive business empires on its successful exploitation. Other companies are beginning to see that they have data (especially on their customers but potentially also their employers or their supply chain) and that they could develop some kind of big data innovation strategy. However, this is not always as straightforward as they might first have anticipated. For instance, Günther et al. (2017) provide an illustrative case study of a European postal service organisation that is active in the business-to-business market, delivering addressed mail for customers such as banks and utility companies. Given the shrinking market for traditional mail services due to digitisation, company executives saw big data initiatives as an opportunity to expand their business notwithstanding. Specifically, they considered selling addresses to their business clients on the basis that their clients would be able to target their advertising based on various household characteristics drawn from the data the company had gathered (e.g., income, family composition, mail received). To advance the intended strategy, the company acquired a data-driven start-up that had been collecting data on socio-demographics, real estate and other household data. Challenges soon emerged, however, and these it found difficult to overcome. These challenges included difficulties in seeing how the acquired capabilities present in the start-up would fit with existing organisational structures, processes and attitudes. Additionally, a major sticking point was that the company did not own the data that it saw as being potentially valuable – contractually, the data belonged to its client companies and could not be sold on. Therefore, in addition to organisational rigidities, there are legal and ethical considerations to be borne in mind with regard to how data can be used. However, as we will see in the next section, the laws are not always effective, and companies have found ways of circumventing legal constraints.

Making digital innovation more ethical

The focus on actual behaviours, rather than trends associated with demographic characteristics that we have no control over, is arguably a 'better' and more ethical form of discrimination for all types of digital innovation, whether this be in relation to marketing campaigns, the development of products and services to cater for particular niches, or government initiatives to promote security or tax collection, for example. This is because we can change our behaviours – like not driving so fast so that we don't have to brake very quickly – while we can't generally change our demographics – we are born male or female, black or white, and most of us stay male or female, black or white for life. As we know, in the past, certain demographic characteristics have been associated with unfair discrimination in relation to job opportunities as well as a whole host of other forms of discrimination such as access to higher education and housing. As a result, many countries have enacted legislation that makes these forms of discrimination illegal, as with various equal pay and equal opportunities legislation (Carey, 2015).

However, this does not mean that there is no downside to this use of personal data to discriminate against individuals, as we have seen. For this reason, it is important that collecting little data – the minutiae of private citizens' lives – should *always* mean that those monitored are aware of it. This awareness, we argue, cannot be dealt with by businesses simply producing very long and complex 'terms and conditions', which only expert lawyers can fully understand, let alone read (Madden, 2012). Businesses aim to collect data in exchange for the new products and services they develop, whether this relates to the possibility to rejoin with former classmates 'for free', by accessing a very expensive digital infrastructure like Facebook, or to benefit from a discount on their yearly car insurance premium. However, we are 'walking data generators' (McAfee et al., 2012) need to be fully aware that our actions will have consequences, and these consequences might well go beyond the ads we will see on social media or a change in the monthly rate to our car insurance, as we have seen with the examples of the filter bubble and technology dependency (Rosen, 2012). We pick up this theme in Chapter 10 when we consider responsible innovation (Figure 9.10).

To provide an example of the kinds of privacy issues that may arise from the use of our data, we can look at the example of Facebook. Facebook conducted a mood experiment in which users' news feeds – positive or negative – were filtered to assess the impacts on users' emotional states (Jouhki et al., 2016). This experiment was undertaken in 2011 without the users' consent, either beforehand or immediately afterwards. When Facebook users become aware of the experiment in 2014, many of them felt extremely upset. This is an example of how implicit connectivity – data captured from social media feeds without the awareness of the data creators – might threaten basic ethical principles such as the right to give informed consent about participation in a research study. Such unethical practices of not providing proper information to study participants such as the Facebook mood experiment – as with other, less well-known social media experiments (e.g., the OKCupid matching experiment – see Hern, 2014) – threaten users' privacy.

Figure 9.10

The Dark Web and the underground internet challenge current laws and regulations
©Cultura

Indeed, most research that utilises big data is generally conducted with very poor security measures that threaten data anonymity (Narayanan and Shmatikov (2008). In Chapter 8, we noted that MTurk study participants have cause to worry because their responses might not be kept anonymous. But that was explicit connectivity because the MTurk 'workers' are at least aware of their participation in research studies. However, they are not necessarily aware of the security measures that Amazon takes to keep their personal data anonymous. With a social media experiment, the ethical implications are more profound because posts/likes and sensor-based data, such as checking in to a hotel, are used as research data. The users remain unaware that their data are being used for a study – or worse – that their data are being manipulated, as happened with the Facebook experiment. Furthermore, the laws and regulations in place are often extremely vague and inconsistent across fields and countries, with study participants running the risk of being 're-identified' (Grimmelmann, 2015). These risks are taken, whether by companies, research institutions, universities or colleges, because it is argued that the innovative findings arising from this type of research are meaningful. It is without doubt that the Facebook mood experiment represented an important research finding because it demonstrated that daily news, whether good or bad, on social media has the ability to condition the psychological state of individuals. However, we believe it is important to question these practices and suggest that data made available from computing applications and digitised objects – especially in the context of implicit connectivity – should be treated with respect to the ethical rights of the actual people who own *their* data (Markus, 2016). We thus begin the final chapter with a consideration of how we can evaluate what is ethical innovation.

Conclusion

The world is increasingly digitised – we connect to people and organisations through computers and we use objects that include sensors and tracking devices that can support our activities. This means that we each leave an increasingly large 'data trail' – data that provide information on what we have done, with whom, where and how. We refer to this as implicit digital connectivity because the residual data arise from the actual activity that we are engaged in and we are not generally aware of the data that we are leaving and how it may be used. But, as we have seen, such data are valuable since companies can develop algorithms that can predict what an individual likes and wants, so that information can be then targeted and personalised, and people can be better supported in what they are doing through the development of innovative products and services. However, we have also seen the downsides of this, including that we increasingly live in a filter bubble, are becoming perhaps over-dependent on a technology 'crutch', have less job security and are losing our right to privacy.

This suggests that there is a need to consider how to encourage more responsible use of data and digital technologies for innovations. According to our analysis, it does not seem realistic to expect that specific laws can fully regulate or enforce what people, whether they are private citizens or business managers, can or can't do in terms of exploiting computing applications and digitised objects. We therefore propose a joint approach where the main actors involved each take responsibility for their *doings*. Opportunities concerning knowledge creation and sharing and digital innovation offered by emerging technologies (e.g., exploring the opportunities of big/little data) need to be pursued. At the same time, mitigating the dark side of datafication should be an imperative for all parties involved.

? DISCUSSION QUESTIONS

The following discussion questions are relevant to using this chapter in teaching exercises and discussions or for revision:

1. Provide examples of implicit digital connectivity creating innovation opportunities.

2. In what ways can implicit digital connectivity create opportunities for companies to improve their marketing strategies?

3. What are some benefits of implicit digital connectivity and innovation for consumers?

4. What is the difference between big data and little data innovation?

5. What are some of the dark sides of implicit digital connectivity innovation?

Case questions

The following case questions might also be relevant to using this chapter in teaching exercises and discussions:

1. What are the main benefits for insurance companies and customers that emerge from the use of the black-box in vehicles?

2. Some innovations relate to new technologies applied to existing business models; other innovations relate more to new ways to use old/mainstream technologies in novel ways. To this end, how do you see the introduction of the black-box in the automotive insurance industry?

3. Monitoring drivers and creating ad hoc premiums based on their driving style clearly constitute an innovative business model. Could this business model based on sensors that monitor customer behaviours be translated into other industries? Provide examples and arguments in answering this question.

4. In the automotive insurance companies case, drivers have to 'sign off' that they are willing to have their driving data collected and analysed in return for a potential reduction in driving insurance premiums. Does this require-ment for users to agree to the black-box overcome all the ethical issues associated with this form of monitoring?

 ## Additional suggested readings

Perdersen, E.R.G., Gwozdz, W., and Hvass, K.K. (2016). Exploring the rela-tionship between Business Model Innovation, Corporate Sustainability, and Organizational Values within the Fashion Industry. *Journal of Business Ethics*.

Stilgoe, J., Owen, R. and Macnaghten, P. (2013). Developing a Framework for Responsible Innovation. *Research Policy*, 42, 9, 1568–1580.

References

Acquisti, A. 2012. "Nudging Privacy: The Behavioral Economics of Personal Information," in *Digital Enlightenment Yearbook 2012*, J. Bus, M. Crompton, M. Hildebrandt and G. Metakides (eds.). Amsterdam, Netherlands: IOS Press, pp. 193–197.

Bode, L. 2012. "Facebooking It to the Polls: A Study in Online Social Networking and Political Behavior," *Journal of Information Technology & Politics* (9:4), pp. 352–369.

Brynjolfsson, E., and McAfee, A. 2014. *The Second Machine Age: Work, Progress, and Prosperity in a Time of Brilliant Technologies*. New York: WW Norton & Company.

Carey, P. 2015. *Data Protection: A Practical Guide to Uk and Eu Law*. Oxford, UK: Oxford University Press.

Clark, C. E., and Newell, S. 2013. "Institutional Work and Complicit Decoupling across the Us Capital Markets: The Work of Rating Agencies," *Business Ethics Quarterly* (23:1), pp. 1–30.

Gerlach, J., Widjaja, T., and Buxmann, P. 2015. "Handle with Care: How Online Social Network Providers' Privacy Policies Impact Users' Information Sharing Behavior," *The Journal of Strategic Information Systems* (24:1), pp. 33–43.

Grimmelmann, J. 2015. "The Law and Ethics of Experiments on Social Media Users," *Colorado Technology Law Journal* (13), pp. 219–272.

Günther, W. A., Rezazade-Mehrizi, M. H., Feldberg, F., and Huysman, M. 2017. "Debating Big Data; a Literature Review on Realizing Value from Big Data," *The Journal of Strategic Information Systems* (26:3), pp. 191–209.

Hay, M., Miklau, G., Jensen, D., Towsley, D., and Weis, P. 2008. "Resisting Structural Re-Identification in Anonymized Social Networks," *Proceedings of the VLDB Endowment*, pp. 102–114.

Hedman, J., Srinivasan, N., and Lindgren, R. 2013. "Digital Traces of Information Systems: Sociomateriality Made Researchable," *International Conference on Information Systems (ICIS)*. Milan, Italy.

Hern, A. (2014). OKCupid: We experiment on users. Everyone does. The Guardian, 29th July. (https://www.theguardian.com/technology/2014/jul/29/okcupid-experiment-human-beings-dating).

Huang, J., Henfridsson, O., Liu, M. J., and Newell, S. 2017. "Growing on Steroids: Rapidly Scaling the User Base of Digital Ventures through Digital Innovation," *MIS Quarterly* (41:1), pp. 301–314.

Jouhki, J., Lauk, E., Penttinen, M., Sormanen, N., and Uskali, T. 2016. "Facebook's Emotional Contagion Experiment as a Challenge to Research Ethics," *Media and Communication* (4:4), pp. 75–85.

Venturini, T., and Latour, B. 2009. "The Social Fabric: Digital Traces and Quali-quantitative Methods". Futur En Seine 2009, May 2009, Paris, France. Cap Digital, Proceedings of Futur En Seine.

Lemaire, J. 2012. *Bonus-Malus Systems in Automobile Insurance*. New York: Springer Science & Business Media.

Madden, M. 2012. "Privacy Management on Social Media Sites," pp. 1–20, PEW Internet Report.

Marabelli, M., Hansen, S., Newell, S., and Frigerio, C. 2017. "The Light and Dark Side of the Black Box: Sensor-Based Technology in the Automotive Industry," *Communication of the AIS* (40:16), pp. 351–374.

Markus, M. L. 2016. "Obstacles on the Road to Corporate Data Responsibility," in *Big Data Is Not a Monolith: Policies, Practices, and Problems*, C.R. Sugimoto, H.R. Ekbia and M. Mattioli (eds.). Cambridge, MA: The MIT Press, pp. 143–162.

Martinez, M. G., and Walton, B. 2014. "The Wisdom of Crowds: The Potential of Online Communities as a Tool for Data Analysis," *Technovation* (34:4), pp. 203–214.

McAfee, A., Brynjolfsson, E., Davenport, T. H., Patil, D., and Barton, D. 2012. "Big Data: The Management Revolution," *Harvard Business Review* (90:10), pp. 61–67.

Narayanan, A., and Shmatikov, V. 2008. "Robust De-Anonymization of Large Sparse Datasets," *Security and Privacy, 2008*: IEEE, pp. 111–125.

Newell, S., and Marabelli, M. 2015. "Strategic Opportunities (and Challenges) of Algorithmic Decision-Making: A Call for Action on the Long-Term Societal Effects of 'Datification'," *The Journal of Strategic Information Systems* (24:1), pp. 3–14.

Nolan, R. L. 2012. "Ubiquitous IT: The Case of the Boeing 787 and Implications for Strategic IT Research," *The Journal of Strategic Information Systems* (21:2), pp. 91–102.

Pariser, E. 2011. *The Filter Bubble: What the Internet Is Hiding from You*. Penguin UK.

Rosen, J. 2012. "The Right to Be Forgotten," *Stanford Law Review Online* (64), p. 88.

Ross, C., Orr, E. S., Sisic, M., Arseneault, J. M., Simmering, M. G., and Orr, R. R. 2009. "Personality and Motivations Associated with Facebook Use," *Computers in Human Behavior* (25:2), pp. 578–586.

Somaiya, R. 2014. *How Facebook is Changing the Way Its Users Consume Journalism*. The New York Times, Available at: https://www.nytimes.com/2014/10/27/business/media/how-facebook-is-changing-the-way-its-users-consume-journalism.html.

Sørensen, C., and Landau, J. S. 2015. "Academic Agility in Digital Innovation Research: The Case of Mobile ICT Publications within Information Systems 2000–2014," *The Journal of Strategic Information Systems* (24:3), pp. 158–170.

Tait, A. 2017. *Will 'dark ads' on Facebook really swing the 2017 general election?* NewStatesman, Available at: https://www.newstatesman.com/politics/june2017/2017/05/will-dark-ads-facebook-really-swing-2017-general-election.

Yoo, Y. 2010. "Computing in Everyday Life: A Call for Research on Experiential Computing," *MIS Quarterly* (34:2), pp. 213–231.

Youyou, W., Kosinski, M., and Stillwell, D. 2015. "Computer-Based Personality Judgments Are More Accurate Than Those Made by Humans," *Proceedings of the National Academy of Sciences* (112:4), pp. 1036–1040.

10 THE FUTURE OF DIGITAL INNOVATION: THE ROLE OF RESPONSIBLE AND FRUGAL INNOVATION

Summary

In this final chapter, we conclude with an outlook on the future of digital innovation. Here, we focus in particular on reviewing the concepts of responsible and frugal innovation by considering what motivates the development of these models of innovation and the potential consequences of their use in both developed and developing countries. Responsible innovations are aimed at innovating for the social benefit of individuals and societies, with an emphasis on sustainability, acceptability and social desirability. Frugal innovation has similarities, but with more focus on innovating with fewer resources and being creative in developing innovations that have a core purpose of doing more with less and changing society for the better. In outlining responsible and frugal innovation, we recognise that at present the two terms are emerging, lack agreed definitions and have some blurring in relation to one another. We also outline some key actions required for its successful implementation and conclude by offering a summation including a comparison between responsible and frugal innovation.

Learning Objectives

The learning objectives for this chapter are to:

1. Consider responsible and frugal innovation as future-oriented models of digital innovation and their difference to traditional models of innovation

2. Apply simple ethical criteria to evaluate whether a digital innovation can be judged to be responsible

3. Examine the core principles and key actions in successful frugal innovation

4. Review key societal and organisational issues associated with digital innovation and the relevance of responsible and frugal innovation to such issues.

Case: Frugal innovation in India

Few substantial case studies exist that examine frugal innovations and why they are coming to prominence, particularly in developing countries. India remains an emerging superpower and is set to become the world's most populated country by 2025. With this comes various challenges, and India faces many issues within its diverse population and huge degrees of wealth and social inequality. This makes India an interesting context in which frugal innovations are thriving and have demonstrable societal, and economic, impact. This is especially true where the wealth divide means that many citizens live in poverty and struggle to afford the prominent innovations (such as in education, technology, healthcare and transportation) common in more developed countries despite the increasing social and economic need for such innovations. While neither example described below is a pure digital innovation, both rely heavily on digital technology for their manufacture and marketing.

The first frugal innovation examined in this case is the Tata Nano. The Nano was introduced into India as an ultra-low-cost, ultra-safe car designed to be accessible to millions of people. However, the Nano encountered numerous issues relating to its low price point, lack of safety features and the controversy it caused relating to potential damage to the environment – through potentially putting many more cars onto India's roads and already polluted cities. This frugal innovation is positioned here as a story of promise, which ultimately led to failure. The second innovation focuses on the Jaipur Foot, a low-cost medical innovation aimed at improving the lives of India's disabled, specifically amputees, who

> typically lack access to the basic medical care and more advanced treatments and prosthetics available in developed countries. This frugal innovation is, in contrast to the Tata Nano, positioned as a long-term success story where an innovation and the concept of 'doing more with less' have produced positive results.

The Tata Nano

The Tata Nano was widely documented as the world's cheapest new car on its launch in 2009 (Dhume, 2011) and was marketed with an initial introduction price of 100,000 Indian Rupees (which was approximately USD $2,000 at the time). It was launched by Tata Motors (an arm of the conglomerate Tata Group) amid much anticipation. The Nano concept car was first unveiled in January 2008 at the Delhi AutoExpo, and production continued throughout 2008 ready for its launch. The principal vision for the Nano, and its focus in this chapter as an example of frugal innovation, came from Tata Group chairman Ratan Tata's vision of providing a safe and comfortable medium of transport to millions of Indian families who use two-wheeled vehicles to travel, often in large, unsafe groups and under extreme weather conditions in summer, winter and during the monsoon (Palepu et al., 2010). This also, to some degree, reflects notions of responsible innovation in that is has a clear purpose which might be desirable for society but is not so desirable in terms of its being a sustainable product.

Two-wheeled vehicles have traditionally been the main medium of transport in India, where they account for 76 per cent of all motor vehicles, as opposed to the 16 per cent share of four-wheeled passenger vehicles (Tiwari and Herstatt, 2012). Despite some initial success and the clear frugal and, to some degree, responsible philosophy behind the Nano that aimed to tackle a prominent problem with India's ever-growing cities and lack of safe infrastructure, post-launch it was widely perceived to have failed in the market. The Nano's sales have so far fallen behind the immensely high expectations created by unprecedented media hype surrounding its development and launch. According to one estimate, the Nano brought Tata Motors worldwide publicity worth USD $220 million, yet press reports after launch widely quoted an executive from a rival motor firm stressing that 'nobody wants to buy the world's cheapest car' (Palepu et al., 2010).

Indeed, sales of the Nano were marred by safety concerns generated through a number of isolated incidents where the vehicles caught fire and also issues encountered relating to the general stigma of being the 'world's cheapest car'. Further concern came from the Nano lacking typical safety features that are common in mass-produced motor vehicles, such as airbags, stock power steering, an ABS (anti-lock braking system) and an engine immobiliser (all innovations that rely on digital technologies). The car also attracted much criticism for potentially having a huge environmental impact if it was to be widely purchased, by putting many more polluting vehicles on

India's already busy roads and into already polluted cities. Chair of the group, Ratan Tata responded to and recognised the challenges with the Nano while launching the Tata Swach, a responsible innovation by Tata which aims to provide low-cost water filtration for developing countries. Tata reiterated the responsible focus of the Nano, stating that while the group was profit-making, the aim was not to create simply the cheapest products (a typical cost-leadership strategy for firms) but instead to reach the largest number of people and, thereby, trying to make a difference (Tata Motors, 2018). Representatives from India's National Innovation Foundation, who have worked extensively to promote grassroot innovations, were quoted as saying: 'People still feel that good technology still comes from abroad' (Malhotra, 2009, p. 2).

For the Nano, despite its intentions as a frugal innovation, quality concerns had a demonstrable negative impact on its success. This is an issue with frugal innovations, in that they strive for low cost, which may have an impact on quality. However, some have, on the other hand, argued that the Nano cannot be regarded a failure, even though its sales figures did not satisfy the forecasts made by marketing analysts nor the expectations of Tata stakeholders (Tiwari and Herstatt, 2012). First, the Society of Indian Automobile Manufacturers revealed that the launch of the Tata Nano succeeded in boosting the market share for India's four-wheeled passenger vehicles. In 2011–2012, the Nano saw a growth of 6 per cent, selling 74,527 units, and also started to meet overseas demand by exporting over 1,000 units in the first six months of the 2011–2012 financial year. Tata also began efforts to address core safety concerns and also improve fuel efficiency to attract new buyers and ease the concerns of environmentalists. However, the numbers of cars produced each year has fallen since launch.

Tata has committed to keep manufacturing the car, with many in the organisation emotionally attached to its concept as a frugal innovation, perceiving it as contributing in a positive way to Indian society. However, it now represents a loss-making model for Tata, and Ratan Tata has expressed regret in marketing the car as the cheapest. Ultimately, although Nano was still in production as of 2018, Tata recognises there is no demand for this frugal innovation and that the product's future is precarious.

The Jaipur Foot

The second innovation focused on in this case is the Jaipur Foot, a prosthetic foot that has reached thousands of underprivileged, disabled people in India with the help of the Bhagwan Mahaveer Viklang Sahayata Samiti (BMVSS). The BMVSS was established in 1975 in Jaipur and is a secular, charitable group that emphasises its non-religious, non-governmental, non-political, non-regional and non-profit charitable status. It was set up to help the physically challenged in India, particularly those who are financially weak and underprivileged. The founder of BMVSS, D.R. Mehta, found his motivation for its inception from suffering a life-threatening road accident that left him facing possible amputation of a leg. Although his limb was ultimately saved, Mehta realised the problems that disabled people have to face in India to be able to be provided with artificial limbs, especially the underprivileged.

BMVSS is most known for helping to create a programme to bring the Jaipur Foot, an example of frugal innovation, to masses of disabled in India. The Jaipur Foot itself was first developed at the SMS Medical College Hospital, Jaipur, in 1968 by a group of eminent orthopaedic surgeons and highly innovative craftsmen who understood the needs of India's poor. The design was developed to meet the sociocultural needs of handicapped people in India, with their particular needs for prostheses that would permit them to squat, sit cross-legged, walk on uneven terrain, work in wet muddy fields and walk without shoes. The Jaipur Foot distinguishes itself from other artificial feet by not having a central keel, which permits mobility in all planes despite being non-articulated. In particular, the dorsiflexion (a term that represents the toes being brought closer towards a person's shin) at the ankle, which is a special feature of the foot, addresses the cultural and lifestyle needs of its target groups.

BMVSS has grown to be the world's largest organisation for the physically handicapped in terms of fitment of artificial limbs, and the Jaipur Foot as a frugal innovation is the driving force behind the organisation's success, built on the desire to help society through the philosophy of doing more with less. This is where we see a key difference compared to the Tata Nano case, where the more-with-less mantra was still the driving force and societal benefits were promoted, but profit was a major factor. BMVSS provides Indians with the Jaipur Foot and other devices such as crutches, callipers, wheelchairs and other ambulatory aids and appliances free of cost. The organisation raises funds primarily through donations (around 60 per cent), with the rest coming through various governmental grants.

The Jaipur Foot has various features that further demonstrate its creativity and frugality as an innovation. It has close to the same range of functions as a 'normal' human foot and, despite using affordable material, it is designed to resemble a normal limb as closely as possible in look and feel. It is also waterproof and can be worn with or without shoes, which helps in challenging terrains. The foot has often been compared to more advanced and costly prosthetics in developed economies such as the USA and in Europe. For example, a comparison of the Jaipur Foot with the SACH (solid ankle cushioned heel) and Seattle Foot was undertaken at the Royal Liverpool Hospital, UK, and the results revealed that the performance of the Jaipur Foot was more natural and closer to the movements of the normal human foot than these alternatives. While the average life of the Jaipur Foot is around three years, the above-knee prosthesis is a highly advanced ischial containment variety, and this helps ensure appropriate bio-mechanical alignment. The cost of manufacturing a Jaipur Foot equates to around USD $30–45, compared to the thousands of dollars that it costs to manufacture advanced prostheses in the USA and Europe. While the main objective of BMVSS and the Jaipur Foot are the physical and socioeconomic rehabilitation of India's disabled, the organisation also conducts frugal innovation and technical research in developing and improving aids and appliances for the physically challenged. Through its workshops, for example, it delivers seminars for dissemination of knowledge and expertise related to the manufacture of such products to help inspire a new generation of frugal innovators.

BMVSS and the Jaipur Foot are an undeniable success story, and the organisation has grown to expand its base across India and increasingly into new international locations. Owing to its performance, the foot has transcended geographical boundaries and is being used by handicapped people in over 40 countries and is now the most widely adopted prosthetic foot in the world. The charity has conducted more than 50 fitment camps in 26 countries and helped set up independent prosthetic fitment centres in Asia, Africa and Latin America. BMVSS has also been given Special Consultative Status with the Economic and Social Council of the United Nations Organisation for its services.

This example of frugal innovation shows that a focus on profit is not necessarily needed for growth and that there is scope for such innovations to grow through their aims to help society and spread knowledge required for frugal innovation. Indeed, CK Prahalad commented on the organisation in his 2004 book as a key example of innovations which seek to help those 'at the bottom of the pyramid' (Prahalad, 2004). Further, the impact of the Jaipur Foot was pertinently described in *Time* magazine in autumn 1997:

'People who live inside the world's many war zones from Afghanistan to Rwanda may never have heard of New York or Paris but they are likely to know a town in Northern India called Jaipur. Jaipur is famous in strife-torn areas as the birthplace of an extraordinary artificial limb known as the Jaipur Foot that has revolutionised life for millions'.

The Jaipur Foot ultimately demonstrates the potential success and longevity of frugal innovation if implemented by enthusiastic, creative pioneers with the right intentions for executing innovation for societal benefit.

Introduction

In Chapter 9, we discussed some of the broad ethical issues related to the increasing presence of digital technology in the products we use and the services that we access. In this chapter, we take this a step further and look at how organisations themselves can become more responsible, and also frugal, in their approach to innovation (digital and non-digital) and as a way of profiting from such innovation. As the concluding chapter to this book, this focus also offers an outlook on the future of digital innovation in and for society and organisations. Relating to our focus on responsible and frugal innovation, we draw on the examples of frugal innovation introduced in the above case and on further examples of responsible and frugal innovation which we introduce throughout the chapter.

Responsible innovation is a relatively new term in the academic and practitioner literature and is associated with more general issues of sustainability and the triple bottom line – that is, a firm's responsibility for considering its social and environmental impacts as well as simply its financial results. Innovations can then also be considered in terms of these three different dimensions rather than simply in terms of whether they can contribute to the firm's bottom line. Building on this concept of responsibility, we also focus part of this chapter on the related concept of frugal innovation. *Frugal innovation* is concerned with how society, particularly emerging economies, are developing innovations that are less complex and therefore less costly to the end user; essentially, it is the notion of 'doing more with less' and is often focused on charitable innovations but can also emphasise profit-making and the bottom line (Prabhu and Radjou, 2015). We saw both sides of this with frugal innovation in the introductory case and to varying degrees of success. However, it is important to stress that we don't want to imply that all frugal innovations with a charitable purpose are likely to succeed while those with a profit motive are likely to fail. Indeed, there are a number of actions necessary for successful frugal, and responsible, innovation which we begin to uncover later in this chapter (Figure 10.1).

Figure 10.1

Frugal innovations in developing countries help in exploiting existing resources with a view to creating new sources of wellbeing
©iStockphoto

We next consider criteria that we might use to assess whether a digital innovation can be judged as responsible or not. Considering the motivations and intentions and the positive and negative consequences of an innovation can provide a useful basis for this evaluation. We then discuss ethical constructs and their applicability to digital innovation processes and explore the notion of frugal innovation in more detail, focusing on its difference from typical models of innovation, which tend to focus on the bottom line. We go on to examine some key principles and actions for achieving frugal innovation, particularly relating to its prominence in developing economies.

Responsible innovation: What is it and why does it matter?

Responsible innovation: A key consideration

Responsibility is an interesting term; on the one hand, it alludes to positive, attributional overtones of ethics and morality; on the other, it considers evaluation and a degree of social and ethical judgement. In the social sciences, including in business and management and, indeed, innovation discourse, responsibility has been a core recurrent theme. The debates on responsibility became more prominent with public controversies related to government reliance on non-renewables, the environmental practices of large firms, and widespread use of digital and surveillance technologies in everyday elements of life, as we have seen in the preceding chapters of this book. Such issues mean that the concept of responsibility is often positioned in relation to risk – as in risk society – and also justice.

One such example of responsibility's increasing prominence in recent years in the academic and practice literatures is that of responsible innovation. There are growing debates on what the term entails and how it could be developed into a framework to guide both research and innovation. It can be argued that these discourses have primarily emerged through examples in developed countries, but increasingly there is attention being placed on the implications it might have in developing countries such as India and China (Crescenzi and Rodriguez-Pose, 2017). The general view of responsible innovation is ultimately that organisations, governments and the public must strive to follow a process that seeks to promote creativity and opportunities for science and innovation that are socially and environmentally desirable and undertaken in the public interest (Figure 10.2).

Figure 10.2

India and China are among these countries where frugal innovation has found fertile ground
©*Getty Images/Stocktrek Images*

How can we judge responsible innovation? Ethical constructs applied to innovation processes

In Chapters 8 and 9, we portrayed the growth of computing applications and digitised objects in somewhat 'Wild West' terms, given the many different organisations attempting to exploit the data trails that we leave to provide more personalised new products and services, purportedly for our benefit. We have also identified in earlier chapters examples of digital innovations that had unintended negative consequences. We have seen, in other words, that not all innovations are 'good for all' in terms of their impact. The question can thus be raised as to what might constitute a way to approach innovation that would make it more responsible. In order to explore this issue, we will build upon our discussion regarding the meaning of responsible innovation and consider how ethical constructs can help us realise a more responsible output from digital innovation efforts.

Ethical judgements are not always 'black and white', so that in many cases innovations cannot be straightforwardly described as either responsible or irresponsible. Nevertheless, we can apply two criteria to weigh up an innovation, digital or otherwise.

A *consequentialist* view considers the positive and negative (happiness and harm) consequences of an innovation's use. A *non-consequentialist* view considers the motivations or intents (especially in relation to the environment and society as well as simply making a profit) that lead to the innovation's development. We expand upon and discuss these in more detail here, although it should be remembered that there is no absolute standard against which to judge a behaviour as being ethical or not, so the following should be seen as guiding principles only:

1. *Non-consequentialism* – This approach to ethics focuses on the intentions rather than the outcomes. In other words, it matters *why* you do something, regardless of the outcome of what you do. For instance, deontology suggests that a person's actions can be assessed against the criteria of duty, with duty sometimes being translated into actions that you or an organisation would be happy for others to follow against you (or your organisation). 'Do as you would be done by' would be another way to put it. In following this approach, innovations would be motivated not only by the desire to make a profit, even in for-profit organisations, but also by a goal or duty to at least leave the social and physical environment no worse and hopefully better than prior to the innovation's appearance. In this sense, taking an ethical or responsible approach to digital innovation from this perspective would mean that the social and environmental issues would weigh as heavily in decision-making as the financial – so innovations that are good for the triple bottom line (Elkington, 1999) would be the motivating drive.

2. *Consequentialism* – This approach to ethics focuses on the outcomes or results: whatever has the 'best' outcome is the ethical choice to make. For instance, within this approach is *utilitarianism* and this suggests that the goal is to maximise happiness and minimise harm. The dilemma of course is how to weigh happiness and harm – if something makes 99 per cent of people happier but the remaining 1 per cent are harmed, does the positive justify the negative? But, in general terms, a decision-maker tries to assess their choices against the likely outcomes and selects the choice that maximises happiness while minimising harm. Of course, sometimes the consequences are unforeseen and unintended, but the decision-maker should at least attempt to evaluate the positive and negative consequences of different innovation options and select the option that is likely to do most good and least harm with respect to social, environmental and financial criteria.

Here, we can see that it would be difficult to defend a decision that did harm, even if the majority were better off or happier as a result of it. Think, for example, of the example of the Ford Pinto, a car based on a strategy to produce a vehicle for the masses that would cost only USD $2,000 in 1970 when the first model was launched. In this example, to save costs (and so make many people happier because they could now afford a new car, as with the Nano discussed in the case above), Ford put the fuel tank in the rear, even though there was a small risk that this meant the car would more easily catch on fire if a collision occurred from the rear. During development, the project team became aware of this defect, but they undertook a cost–benefit analysis of what the likely risk and costs of death/serious burn accidents would be as against the added cost of changing the design. Based on this analysis – which included the costs of paying out to those who suffered from any crash – it was decided to leave the fuel tank where it was, because moving the tank would delay the release and increase costs. As it turned out, the company was eventually charged with reckless homicide following a number of fatal accidents. It was eventually acquitted of the charges but not after serious reputational damage and paying out USD $128 million to victims – a sum considerably higher than that which it had predicted in the earlier cost–benefit analysis. More importantly, from our consideration of ethics, we could argue that the focus on analysing duty rather than consequences might have reduced the chances of the company going ahead with production once it was aware of the design defect.

The Ford Pinto case is a clear example of irresponsible innovation. In practice, taking a responsible innovation approach is probably most likely when an organisation starts from the position that it wants to 'do no harm' and/or 'leave the world a better place' and then combines this with a thorough analysis of the potential consequences of any innovation it is thinking of bringing to market – social and environmental as well as financial. Unfortunately, many organisations include strong statements about their duty to the environment and society while not always seeming to weigh these elements as heavily in their analysis of their innovation projects as they do the potential financial benefits. Of course, in a competitive environment, maintaining a profit (or for not-for-profit organisations, sufficient money to cover operational costs) is important, but this need not mean that the social and environmental consequences of an innovation project are not given equal consideration.

This is evident when considering some of the problems associated with digital innovation, which is a key theme and central to many of the examples that we have covered in this book. We can see that, in these examples, both criteria are not fully met by some of the current strategies being adopted. For example, from a consequentialist perspective, developing algorithms so that people more easily access things that interest them is likely to have at least some positive consequences, with many people being happy with this filtering. However, companies have not really considered any harm that they may do due to people becoming exposed to a more limited range of ideas and information, which may lead to increasing levels and instances of intolerance. There are also the consequences of people being able to access unsavoury content on the internet – whether, for example, these are inappropriate images of children being shared or terrorist propaganda accessed by people who become radicalised. Arguably, these harms have not been sufficiently considered in the policies and filtering criteria of internet firms. From a non-consequentialist perspective, one could argue that it is a company's duty to understand the impact of any such filtering. From both perspectives, therefore, one can argue that, currently, companies' activities with digital innovations are not always being motivated by a desire to leave the world a better place and are causing at least some harm. As a result, it would be difficult to evaluate such digital innovations as responsible from an ethical perspective.

Key opportunities and challenges for responsible digital innovation

Following our discussion of responsible innovation and its relation to key ethical considerations, we explore the key opportunities and challenges for this model of innovation in relation to governance and research. Indeed, it is in relation to governance and research where this emerging concept of responsible innovation has perhaps been most discussed and is making noticeable impact. This new push for responsible innovation has some of its roots in debates about the responsible use of emerging technologies, such as nanotechnology (around 2007), and reflections on how this is interlinked with established frameworks of ethics, governance, public engagement and research.

Responsible innovation is becoming a new model of innovation but is also becoming central to how we think about governance and research, particularly for considering relations between innovation and society. For example, as responsible innovation is being used and discussed ever more frequently in the USA, Europe and further afield, it is beginning to play a role in both regulatory and funding institutions. This is observable through the European Union's Horizon 2020 research funding project, for example, where projects are expected to demonstrate responsible innovation (European Commission, 2018). Similar guidelines are also beginning to come to prominence in the USA, with new funding schemes focusing on integrating aspects of the responsible innovation concept (e.g., in the Office of the Comptroller of the Currency). Funders and policy-makers alike hope that, through responsible innovation, new developments produced from research can arise on a moral basis and that the concept can steer innovation models towards making desirable impacts in an ethical and considered way.

In principle, responsible innovation seems to be a scheme for good and offers various opportunities for society generally and for research and development relating to innovation. However, this has brought some challenges for responsible innovation. For instance, it is important to clarify its role in organisations, governance and research implications and, in particular, whether this is sustainable in the long term. We contrast the key positives and opportunities discussed here and earlier in the chapter with a number of potential challenges for responsible innovation as a future-oriented model of innovating.

Challenges in governance and research

A first challenge is that there is much potential for responsible innovation to become something of a buzzword for organisations, policy-makers and in research – something without much substance or proof of application. There are limited guidelines or agreed methods for judging or measuring responsible innovation despite our promising discussion of this in the previous section. Equally, demonstrating clear impact of responsible innovation might be difficult depending on the context, industry or project. This means that if organisations, policy-makers and researchers are pressured into demonstrating responsible innovation, we might begin to see discontent towards the notion, with people trying to find ways of working around systems to display this and, ultimately, the concept losing some legitimacy.

A second challenge relates to clarity. If we are still unsure what responsible innovation means (as in there are no clear definitions or agreed principles), then how can policy-makers and researchers be expected to integrate it in their regulations and projects, and how do they go about doing this? These are important questions, especially as the notion of responsible innovation has begun to have influence in this regard relating to governance and research. For example, funding agencies in the European Union increasingly require responsible innovation processes to be embedded within their funded projects.

Third, we have to consider what the term 'responsible innovation' means and how this might bring about misconceptions regarding typical models of innovation. Indeed, does this mean innovation as we more traditionally know it is now deemed irresponsible? While we may all have ideas of what irresponsible innovation might be according to our personal convictions and beliefs, deciding on what responsible innovation would be for society as a whole might prove significantly more difficult. Of course, innovation always has unforeseen consequences and is full of risk and uncertainty, which has historically led to the deployment of regulation aimed to manage such risks and uncertainties, such as in the control and strict regulation of medical innovation and nuclear engineering. However, as we saw with the Tata Nano, there can be conflicting aspects of responsibility – in this case, between the social benefits of the car versus the environmental problems that increased car ownership brings. Despite these caveats, thinking about duty to society and the environment, by including a thorough review of the broad range of

Figure 10.3

Nuclear plants, and associated innovation, have been hugely regulated in the past decades
©Getty Images/iStockphoto

potential consequences to both, would certainly be a start in terms of avoiding some of the more irresponsible innovations, such as the Ford Pinto, that have occurred over time (Figure 10.3).

Frugal innovation: An emerging consideration in developing and developed economies

As in our introduction about responsible innovation, here we start by considering the nature of the word 'frugal' and its increasingly prominent application to innovation. The concept of frugality historically points to the quality of being sparing, thrifty, prudent or economical in the consumption of key resources. This might include food, time or money and also emphasises avoiding waste and extravagance (Lastovicka et al., 1999). Frugal innovation is therefore the concept of 'doing more with less'. It is more than an innovation model or strategy and instead denotes a new frame of mind that sees resource constraints not as a liability but as an opportunity. It also favours agility over efficiency: frugal organisations don't seek to entice customers with technically sophisticated products but instead strive to create good-quality solutions that deliver the greatest value to customers at the lowest cost, often free of charge, as in the Jaipur Foot innovation case. Much of the current academic discussion on frugal innovation finds its roots in developing and emerging economies, particularly India, as was the focus of the cases that opened this chapter.

However, more and more business leaders and researchers find that such frugal innovation is important for developed economies such as the USA and Europe. This springs partly from a concern that, unless they develop skills for frugal innovation, organisations in developed economies might miss out on the growth of prominent emerging middle-class markets in countries like India and China. On their own turf, moreover, the likes of the USA and Europe are facing growing competition from emerging market competitors that are entering their markets with affordable yet increasingly reliable propositions. Further, the frugal model of innovation might offer new opportunities to make the most of technological expertise, to address societal challenges in developed economies and to better meet customer needs in home markets at a time when consumers strive for value.

In India, to relate again to the context of the cases with which we opened this chapter, these kinds of frugal solutions are called *Jugaad*, which is a Hindi word that means an improvised fix – a clever solution born in adversity. To Radjou (2017, p. 1), the individuals who devise such innovations are 'like alchemists: they can magically transform adversity into opportunity, turning something of low value into something of high value. They're masters of the art of doing more with less.' As we explored with the example of the Jaipur Foot in India, when commodities are scarce, people are forced to look within themselves to tap that most abundant-of-all-natural resources, human ingenuity, and use this to solve issues that can otherwise plague society (Radjou, 2017). As with the Jaipur Foot, frugal innovations can be low-tech, especially when compared in that case with advanced prosthetics in regions such as the USA, Europe and Australia, where there is much higher investment and access to medical innovations. However, frugal innovations can also focus more on utilising the advancements of digital solutions to make services more affordable and more accessible to more and more customers. This came through to some degree with the Tata Nano, where there were some key societal motivations – central to frugal innovation – even though in this case profit was emphasised (Prahalad and Hammond, 2002). Even the Ford Pinto car case could be described as being frugal since the aim was to produce a family car for less so that it would be affordable to more people.

Frugal innovation and established models of innovation

In introducing the concept of frugal innovation, we soon come to realise that it is vastly different from, some might say opposed to, the established models of innovation that we see in developed countries. For example, in California, which in 2018 overtook the UK to become the world's fifth largest economy behind the USA as a whole, China, Japan and Germany, giant tech firms in Silicon Valley focus on chasing the 'next big thing'. Companies spend billions of dollars investing in R&D and create ever more complex digital products. For instance, think again of the example we introduced in Chapter 2 about incremental innovation and the iPhone with its multiple generations. For Apple, this is a classic example of differentiating its brands from the competition and in turn charging customers more money so that they are seen to be up-to-date in benefitting from the new features being introduced. This has, in contrast to frugal innovation's 'more with less' mantra, been referred to as the conventional business model culture of 'more for more'. Radjou (2017) emphasises that this 'more for more' way of thinking is becoming somewhat obsolete for three main reasons. First, some customers can no longer afford expensive products, due to diminishing purchasing power. However, it is also important to consider that this is only true to some degree and in other cases purchasing power may well have increased. Second, we are running out of natural resources, such as non-renewables, like oil, and fresh water, especially as temperatures rise and droughts become more commonplace. Third, and most important, the growing income disparity between the wealthy and everyone else has led to a big disconnect between existing products and services and customers' essential needs.

To many, a shift from the more-for-more model is needed in developed countries, so they too can start to see the benefits of frugal innovation and begin to move away from relying on more-for-more innovating, especially when the 'more' is perceived as being not really necessary. In the USA and Europe, we have seen glimpses of a frugal innovation revolution led by determined and forward-thinking entrepreneurs with revolutionary solutions to common problems. To refer again to Silicon Valley, the start-up gThrive is a good example. gThrive makes wireless sensors that farmers can place in various parts of their fields with a view to collecting detailed information about soil conditions. The data allow those growers at the bottom of the chain to optimise their use of water while improving the quality of their products and yields, thereby helping to maximise their output. This relatively simple digital innovation helps farmers directly, rather than the stakeholders further down the supply chain who typically benefit the most. Further, in the context of California, which is increasingly faced with wildfires and major water shortages, these digital devices are particularly important and can typically pay for themselves through core saving within one year. In Europe, there have also been calls for more focus on frugal innovation. Qarnot Computing, a company based in France, is an interesting example and has developed a way to use waste heat from computer processors and servers to be recycled for conventional heating, both in businesses and for domestic customers. In doing so, Qarnot aims to distribute the creation of surplus heat to people's homes through specially designed, affordable radiators that are part of the internet of things, being connected to the internet to increase digital

Figure 10.4

Innovation initiatives in Silicon Valley include those addressing sustainability
©Getty Images

connectivity, interaction and efficiency. The innovation is also particularly useful for businesses, as at present around 2 per cent of global electricity use is attributed to servers. This is also an example of the *circular economy*, where waste from one product or process is used by another product or process (Figure 10.4).

Three principles for frugal innovation

While we have briefly discussed the difference between frugal and other models of innovation, it is also important to clearly outline its key principles. We introduce three principles here which focus on frugal innovation's simplicity, creativity and agility (Radjou, 2017):

1) The first principle is to ensure simplicity, avoiding innovations that are designed with advanced, often 'luxury' features aimed at impressing consumers and users. Key to frugal innovation is to make products and services easy to use, widely accessible and with a clear purpose for individuals and society. Again, we can discuss this in comparison to traditional models of innovation where over-engineering of complex products in insular R&D labs is commonplace. In frugal innovation, innovators are often in the field, observing customers in their natural settings to identify their real-life challenges. In research, this also has an interesting connection with ethnographic modes of collecting empirical data; guided by a rich understanding of situations, frugal innovators can zero in on features that

could solve pressing individual and societal problems. In the case of the Jaipur Foot, innovators recognised the problem from spending time with disadvantaged communities in India and continue to work closely with people to understand their disabilities and increasingly diverse needs.

2) The second principle is creativity, particularly through leveraging existing resources and assets that are readily available. The materials used for the Jaipur Foot innovation demonstrated this, particularly in showing how creative use of simple materials (especially when compared with advanced materials used in prosthetics in more developed countries) can yield outstanding results in certain contexts faced with specific needs. Frugal innovators can also borrow proven technologies in one sector and adapt them to make new products in a different industry or combine and integrate multiple existing technologies to create a new frugal innovation. Taking another example from India to complement the cases introduced at the beginning of this chapter, GE Healthcare developed a low-cost, portable electrocardiogram (ECG) device, the MAC 400, which uses established technologies but adapts these to be cheap and suitable enough to operate in harsh conditions in rural areas. The printer for the MAC 400, for example, was adapted by GE Healthcare's innovation team from a printer typically used to print bus tickets.

3) The third principle is to be agile. While companies following traditional models of innovation tend to scale up vertically by centralising operations in large-scale factories and warehouses, frugal innovation focuses more on thinking and acting horizontally. Indeed, this represents more focus on being agile and scaling out horizontally through a supply chain which emphasises smaller-scale manufacturing and distribution. The same 'downsizing' logic can also apply to distribution, where large 'brick and mortar' retail stores can be replaced by smaller stores or online distribution channels. Across the Philippines, for example, sari-sari stores are small family-run shops located in hundreds of villages. Hapinoy, a social enterprise launched by MicroVentures, provides specific training and financing to owners of sari-sari stores so they can upgrade their businesses to offer further services. This is also relevant to the Taobao villages example discussed in Chapter 8.

🗨 Key Concepts: Responsible and Frugal Innovation

Responsible Innovation:
A model of innovation that adopts the core principles of transparency and acceptability, enabling a process where innovators strive to innovate in an ethical, sustainable and desirable way to meet social and environmental goals as well as financial goals

Frugal Innovation:
The concept of 'doing more with less' in innovation, where innovators consider societal and economic needs as the main driver for innovation. Achieving this, particularly with the aim of developing marketable products for developing markets, occurs through a process of reducing the complexity and cost of innovations, such as by removing non-essential features

The challenges of frugal innovation

It is also important to emphasise key challenges that might face frugal innovators. One such challenge is the potential resource constraints that are typical in developing countries and economies. Consumers and users in these markets are very value conscious and, indeed, many have only recently shifted from being non-consumers to consumers with some buying power. An increasing segment of these markets is made up of an emerging middle class whose members live on approximately USD $2–13 per day. This middle class is a rapidly growing segment of the population. In India, 5 per cent of households are predicted to earn around USD $4,000 per annum by 2020, which is a substantially higher percentage than seen in previous decades (Williamson and Zeng, 2009). Despite their increasing purchasing power, these new consumer households will still have limited disposable income and will remain conscious of getting maximum value for money on their purchases and investments. Consumers in Western markets are also becoming more value-oriented, especially in the wake of the 2008 financial crisis. An example of this has can be seen in supermarkets in the UK, where emerging no-frills, value-oriented players Aldi and Lidl have caused problems and profit shocks for the 'big four' (Asda, Morrisons, Sainsbury's and Tesco). The proposed merger between Asda and Sainsbury's is further proof that these established players are exploring 'frugal' avenues to maximise profit potential and to make necessary savings throughout their supply chains by using digital solutions. The challenges for developed countries relating to frugal innovation, therefore, are interesting and highly significant.

Many frugal innovations are simply not needed in some countries, and so recognition of the concept is challenging for organisations and individuals unless there is a clear need and societal benefit. An example of this is the Mitticool, a fridge launched in 2006 in India and made entirely from clay. The fridge uses the concept of cooling through evaporation and therefore doesn't require use of electricity, helping those Indians, for whom a conventional fridge is out of reach financially, to have a safe and reliable place to store fresh food. In developed countries, there is simply no widespread need for this, as fridges are widely available and affordable – even though conventional fridges are much less environmentally friendly as compared to the Mitticool fridge. The point is that multinational companies are unlikely to even think about developing non-electricity-based fridges; rather, they are experimenting with and beginning to market digital fridges that record and alert users about food stocks that need replenishing. This again hints that for long-term success in frugal innovation, developed countries and their corporations must rethink their established models of innovation if frugal innovation is going to become more broadly applied. Typically, the business models of such multinational companies, even when operating in developing countries, have not considered the resource-constrained consumer and the limits of the earth's resources but rather have focused on the affluent few at the top of the pyramid who possess the buying power to afford advanced products such as a digitised fridge.

However, this strategy of earning high margins from a few affluent consumers is increasingly questionable. On the one hand, frugal innovations will over time also attract affluent consumers who may decide to go for less expensive products that still meet their needs. As we mentioned previously, many consumers – even those with huge buying power – are increasingly striving for value. On the other hand, the growing middle class in developing countries is becoming an increasingly interesting market that offers great business potential. If large organisations continue to ignore the growing middle class in emerging markets, they risk losing market share to the rising competition, in both emerging markets and indeed their home markets. These trends might force organisations to adapt, fundamentally altering the way in which products and markets are defined and paired. Second, companies that want to engage in frugal innovation must build organisational structures and capabilities to enable the development of frugal products. As with the business models that shape them, existing R&D processes and structures are often based in large centres and optimised for the development of advanced products and digital technologies targeted at high-end consumers. As discussed earlier in this chapter, frugal innovation takes place among the people who need it most and is often driven by entrepreneurs and creative individuals or groups who are working with local populations. The role of local R&D units, however, has traditionally been to adapt centrally developed products to local needs, thereby exploiting existing competencies and avoiding the costs of new product development (Gassmann and von Zedtwitz, 1998). The resulting products often entail changes in design and the ranges of features, but they are still usually based on an advanced product architecture aimed at consumers in wealthier countries. Conversely, prior research suggests that successful frugal innovation requires a deep understanding of the specific environment for which such products are developed. Having local presence and a fundamental new-product development may therefore be necessary to develop a truly effective frugal innovation process (Williamson, 2010) (Figure 10.5).

Figure 10.5

Creativity is key to frugal innovation
©GETTY

Key actions for successful responsible and frugal innovation

While much of the discussion of responsible and frugal innovation in this chapter has focused on developing countries (where, it can be argued, it is most prominent), and India in particular, it is also important to consider key actions for successful responsible and frugal innovation beyond developing countries and in relation to organisations, policy-makers and research more generally. This means that we can begin to synthesise our discussion of both of these emerging models of innovation and consider the two together in relation to key actions for success. We move on to exploring some additional key areas of applicability for both responsible and frugal innovation to conclude this chapter and guide further understanding of where these innovations are perhaps most prominent.

Five important actions

One prominent text on responsible and frugal innovation by Prabhu and Radjou (2015) studied the leaders of over 50 companies in the USA and Europe at the forefront of these emerging means of innovating. The leaders who contributed to this project offer examples of not only radical adjustment of business models but also changing the knowledge processes and culture of employees in order to weave responsibility and frugality into the corporate mentality. From these examples, Prabhu and Radjou identify five valuable actions that help drive successful responsible and frugal innovation into the knowledge base and culture of an organisation. Next, we discuss these and draw comparisons with the cases in this chapter and other prominent examples. This offers a conclusion to our discussion of both responsible and frugal innovation, building on a number of key discussion points in this chapter.

1) Develop 'circular value networks'. Most companies today operate linear value chains, in which products are designed, produced, sold and then consumed. Ultimately, the products end up in landfills. This linear economic model is wasteful, costly and environmentally unsustainable, and we have seen increasing pressures on organisations in recent years (e.g., in relation to single-use plastics, which end up in the world's oceans, or discarded digital products that end up in landfills creating toxic waste) to stop this waste. To be resource-efficient, organisations must look to frugal innovation and reinvent their value chains to operate in a 'circular' way by embracing new sustainable methods of design and production and distribution. These circular economy techniques enable the continual reuse of materials, parts and components, and even waste (Esposito et al., 2018).

2) Crowdsource solutions across industries. Businesses in the developed world now face 'wicked' problems that are complex and messy. They rely primarily on in-house capabilities and resources, which is increasingly becoming a non-viable strategy for innovation. Instead of wasting time and money trying to reinvent the wheel, organisations must strive to be more cost-effective, and utilise digitally facilitated crowdsourcing for responsible and frugal solutions from external networks of suppliers, research institutions and creative entrepreneurs. Indeed, partnering with external innovators can shift the perspective of employees and speed up cultural change (Majchrzak and Malhotra, 2013).

3) Encourage simplification of structures to empower employees. Companies need to be responsible and frugal with time, one of the most valuable resources in business. To save time and gain agility, companies can learn to fully utilise their assets. This doesn't mean only physical or service assets but also human assets and their knowledge. To address customer needs faster and better, an organisation can simplify organisational structures by eliminating bureaucracy, empowering employees, cultivating a flexible but responsible mindset in its workforce while making the most out of emerging digital technologies. We explored elements of this in Chapter 6, with the discussion of project liminality and open innovation,

and the case of IBM InnovationJams and the city of Vienna (Morton et al., 2018).

4) Use key performance indicators (KPIs) to incentivise and sustain responsible and frugal behaviour across organisations. To elaborate, it is the responsibility of organisational leaders to create specific KPIs to drive both responsible and frugal thinking and action at all levels. These KPIs should enable employees to track performance against enterprise-wide goals and adjust individual and collective efforts. The French food group Danone applies Danprint, a measurement tool co-developed with software provider SAP, to track the carbon footprint of an entire product life cycle in all Danone subsidiaries. Danprint is used to encourage senior managers to become resource-efficient and meet chairman Franck Riboud's goal to cut Danone's carbon footprint by more than 50 per cent by 2020.

5) Emphasise the 'doing more with less' mantra of frugal innovation, in particular, but also the societal and environmentally desirable aspects central to responsible innovation. Organisations should not merely communicate their bold responsible and frugal innovation goals by issuing press releases but should instead put their personal reputations on the line by making major public announcements about these goals and restating them incessantly to employees, customers, investors and partners. Paul Polman, the CEO of Unilever, has done precisely this through the launch of Unilever's Sustainable Living Plan in 2010. Here, Polman publicly committed himself and his company to an ambitious target of doubling the company's revenues while halving its environmental footprint by 2020 (Whittington et al., 2015).

Creating a responsible and frugal innovation culture ultimately requires systemic change across organisations and their knowledge processes, and leaders must lead from the front in initiating such change. Outlining the key actions for success helps in further understanding what is required to drive these models of innovation for organisations, policy-makers and researchers alike. In this sense, we clearly see how organisations will be successful in relation to responsible and/or frugal innovation only if they have a clear declaration of a moral duty to care for the social and physical environment. Further, they must recognise that changes have to be made to reduce negative consequences that their products and services might have. While not all responsible and frugal innovations will be digital, digital technologies will be a fundamental part of many such innovations and digital technologies will also play a major role in their development and diffusion. This means that digital innovation can significantly help in building a more sustainable future as long as those involved recognise the broader ethical and environmental issues of the technologies they are developing and using.

A summary of responsible and frugal innovation: Key areas of applicability

We finish our discussion by looking at a number of areas where responsible and frugal models of innovation are already making key impacts. Many of these areas have an emphasis on digital innovation and involve a variety of industries; they are useful for summarising and bringing together the concepts as we conclude the book and consider some prominent areas of applicability, which form diverse examples of the different models of innovation. This is also useful in illustrating several of the key concepts relevant to this chapter while helping to demonstrate their use in further real-life settings (in addition to the case examples on frugal innovation in India which opened this chapter).

We also look back at the discussion of consequentialist and non-consequentialist perspectives in our first example; readers can consider the other examples in relation to these concepts as well as the central responsible and frugal innovation concepts. This helps to illustrate that it is not always straightforward to decide when an innovation can be classed as responsible as opposed to irresponsible or frugal as opposed to more 'typical' in the innovation process (Prabhu and Radjou, 2015). We encourage readers to be critical in their own analysis of the examples we provide and also to consider other areas of responsible and frugal innovation that they are aware of and are of personal interest. The examples only touch upon the possible areas of applicability; however, those that follow can be used to discuss and debate the applicability of responsible and frugal innovations concepts that have been outlined in this chapter:

1) **Healthcare:** A report by McKinsey (Kayyali et al., 2013) entitled 'The Big-Data Revolution in US Healthcare: Accelerating Value and Innovation' estimated that the US healthcare industry (representing one fifth of the country's economy) will save up to USD \$400 billion by studying aggregated data that are currently being stored on local hospital-based electronic medical records. By connecting various healthcare databases, it would be possible to advance doctors' ability to diagnose diseases and would help researchers understand the root causes of various illnesses/diseases. In addition, the 2015 worldwide implementation of ICD-10 (a standard system providing codes to classify diseases, symptoms and causes of diseases) opened up possibilities to further apply big data research to healthcare. From a non-consequentialist perspective, this can be considered positive as it is motivated by a desire to better understand illness so that the illness can be prevented or treatments better targeted. From a consequentialist perspective, it may be more dubious because the risk of an individual's data being accessed is a threat both to the individual and to society, given the value of such data to the criminal underworld. Such innovations can be considered responsible as they are

clearly aimed at providing healthcare benefits across society. However, they are not really frugal as the infrastructure investment to collect these data is huge.

2) **Utilities:** The State of California uses advanced data analysis platforms for situational intelligence and powerful visualisation and modelling tools to project supply–demand dynamics across the power grid. Real-time visibility is said to be critical because this reflects when and how people expect to get their electricity. The essential database is huge and diverse, including weather forecasts, real-time sensor data and continuous metering. The aim is to reach an optimal balance between supply and demand, plug into renewables when necessary and avoid blackouts and service dips. This will help to conserve energy production and maximise the use of renewable energy sources, with obvious positive effects in terms of the environment.

3) **Retail:** A significant proportion of our shopping these days occurs online. Even traditional brick-and-mortar stores now have an online presence. This trend has consequences for our high streets, for distribution channels, for jobs and for transport systems. We have also seen how companies are using digital trace data to personalise marketing to induce shoppers to purchase particular products and services. It is interesting to debate the different digital solutions in terms of what might be ethically the best way forward for retailing and whether different innovation approaches are responsible and/or frugal. It is also interesting to debate what governments might do to make the retail industry more responsible.

4) **Security:** Automated information collection and intelligence sharing allow smarter links between myriad security systems and layers (including physical protections), enabling continuous and real-time data monitoring combined with behavioural signals (e.g., how people use websites or access servers), and cybersecurity experts to identify potential attacks. So-called security analytics allow systems to automatically adjust their risk profile (i.e., go on high alert) once any system in the 'threat intelligence network' detects a threat, be it malware, a rogue peripheral or suspicious log activity. These security systems rely on increasing use of surveillance technology setting up a tension between individual privacy and societal security. It is important to discuss the tensions here in relation to responsible innovation goals and thus the ethics involved in this kind of approach.

5) **Transport:** There are many innovations being worked on right now that focus on different aspects of our transport system – battery design that can improve the distance travelled in an electric car; evacuated tube transportation technologies that will shoot people and goods in capsules that move much faster and more efficiently than other ground transport

systems through tubes; aircraft design using lighter materials to reduce energy consumption; electric bikes that encourage more individual commuting; and autonomous vehicles that can increase safety and efficiency. All these different transport innovations have at their heart the aim of lessening our reliance on fossil fuels and so reducing air pollution, which is an increasing problem in large cities and urban areas, while improving efficiency and safety. Discussing the different options in terms of the ethical considerations and whether each is responsible and/or frugal will help to identify the opportunities as well as the challenges of these futuristic scenarios for our transport systems.

6) **Social care:** An aging population is a growing problem in many countries. As people get older, they can struggle with the daily chores of life and are more prone to falls and illnesses. While we have identified the risks associated with becoming overly dependent on digitised objects in Chapter 9, where a person begins to struggle with maintaining their independence, such a crutch might be profoundly welcome. Homes can be equipped with motion sensors that provide an alert if a person does not move about in their 'normal' way, helping to provide quick detection against a fall, for instance. Digitised medicine dispensers can help ensure that medications are taken when required and in the correct dosage and this information can be fed back to a doctor. An online link can provide a consultation with a GP so that there is no need to go to the surgery, which may prove difficult for some elderly people. These innovations to allow people to remain independent and in their home for longer can have significant benefits for the individuals and for societies, given the cost of social care. However, a discussion of more frugal approaches to the problems of an aging population might lead to other innovations that might also counter such problems as loneliness and social isolation, for example.

In all these areas of digital innovation, there is evidence of motivations that include improving the environment or society and some very positive consequences for individuals and potentially society, even while in some cases there is an underpinning profit motive. This profit motive is an interesting consideration in contrast to some concepts in frugal innovation, where profit is not so clearly a motive and the societal benefits really come to the forefront in driving innovation. Moreover, in each area outlined above, there is some potential harm involved – even if this is to a minority only (Crescenzi and Rodriguez-Pose, 2017). Thinking about these issues and consequences can allow us to better evaluate digital innovations, helping to identify those that are more responsible and/or frugal and so more likely to encourage global sustainability.

Conclusions

In this chapter, we have discussed the ethics of innovation and considered both responsible and frugal innovation, the differences between these concepts, and why they are important as emerging models of innovation, especially with regard to digital innovations, in organisations and wider society. We recognise that the concepts of responsible and frugal innovation remain somewhat ambiguous. Lacking still are clear definitions of these models of innovation or agreed principles. Equally, there is some blurring between the two concepts, which in many regards represent similar themes in innovation, particularly around innovating for sustainability for social, environmental and economic benefit. However, our aim here has been to discuss responsible and frugal innovation and offer some key examples in relation to possible future areas of digital innovation.

? DISCUSSION QUESTIONS

The following discussion questions are relevant to using this chapter in teaching exercises and discussions or for revision:

1. Why is ethics such an important issue to consider in relation to digital innovation processes?

2. What are the key differences between responsible and frugal innovation? Create a table contrasting the two approaches and give examples of digital innovation to illustrate.

3. Explain the key challenges of both responsible and frugal innovation. How might organisations overcome these challenges? Use examples of digital innovation to illustrate the challenges.

4. Why might frugal innovation be more relevant (at present) in developing countries and less so in developed countries?

Case questions

The following case questions might also be relevant to using this chapter in teaching exercises and discussions:

1. What are the ethical issues in the Tata Nano case? What are the key reasons for the failure of the Tata Nano car? Discuss possible ways Tata could have been more successful in marketing its frugal innovation using digital technology.

2. The Jaipur Foot is seen as a success story relating to frugal innovation. Create a frugal innovation strategy for a digital innovation project of your choosing by drawing on the success of the Jaipur Foot and consider the relevance of the five key actions for frugal innovation.

Additional suggested readings

Crescenzi, R., and Rodriguez-Pose, A. 2017. 'The Geography of Innovation in China and India,' *International Journal of Urban and Regional Research* (41:6), pp. 1010–1027.

Prabhu, J., and Radjou, N. 2015. *Frugal Innovation: How to Do More with Less.* London, UK: Profile books.

References

Bhagwan Mahaveer Viklang Sahayata Samiti Website. 2018. *Jaipur Foot.* Bhagwan Mahaveer Viklang Sahayata Samiti (BMVSS), Available at: http://jaipurfoot.org/how_we_do/technology.html.

Crescenzi, R., and Rodriguez-Pose, A. 2017. 'The Geography of Innovation in China and India,' *International Journal of Urban and Regional Research* (41:6), pp. 1010–1027.

Dhume, S. 2011. *Unloved at Any Speed.* Foreign Policy, Available at: https://foreignpolicy.com/2011/10/07/unloved-at-any-speed.

Elkington, J. (1999). Cannibals with forks: the triple bottom line of 21st century business. Oxford: Capstone.

Esposito, M., Tse, T., and Soufani, K. 2018. 'Introducing a Circular Economy: New Thinking with Managerial and Policy Implications,' *California Management Review* (60:3), pp. 5–19.

European Commission. 2018. *Horizon 2020.* European Commission, Available at: https://ec.europa.eu/programmes/horizon2020/en/ /

Gassmann, O., and and von Zedtwitz, M. 1998. 'Organization of Industrial R&D on a Global Scale,' R&D Management (28:3), pp. 147–161.

Kayyali, B., Knott, D., and Van Kuiken, S. 2013. The big-data revolution in US health-care: Accelerating value and innovation. McKinsey & Company, Available at: https://www.mckinsey.com/industries/healthcare-systems-and-services/our-insights/the-big-data-revolution-in-us-health-care.

Lastovicka, J.L., Bettencourt, L.A., Hughner, R.S., and Kuntze, R.J. 1999. 'Lifestyle of the Tight and Frugal: Theory and Measurement,' Journal of Consumer Research (26:1), pp. 85–98.

Majchrzak, A., and Malhotra, A. 2013. 'Towards an Information Systems Perspective and Research Agenda on Crowdsourcing for Innovation,'. *The Journal of Strategic Information Systems* (22:4), pp. 257–268.

Malhotra, J. (2009), "Questions and Answers: Prof. Anil Kumar Gupta", Wall Street Journal, 24.09.2009.

Morton, J., Stacey, P., and Mohn, M. 2018. 'Building and Maintaining Strategic Agility: An Agenda and Framework for Executive IT Leaders,' *California Management Review* (61:1), pp. 94–113.

Palepu, K.G., Bharat, A.N., and Tahilyani, R. 2010. *Tata Nano – The People's Car.* Harvard Business School Case, Available at: https://www.hbs.edu/faculty/Pages/item.aspx?num=38716.

Prabhu, J., and Radjou, N. 2015. *Frugal Innovation: How to Do More with Less.* London, UK: Profile books.

Prahalad, C.K. 2004. 'Co-creating unique value with customers,' *Strategy & Leadership* (32:3), pp. 4–9.

Prahalad, C.K., and Hammond, A. 2002. 'Serving the World's Poor, Profitably,' *Harvard Business Review*. September.

Radjou, N. 2017. *The genius of frugal innovation*. TED Ideas, Available at: https://ideas.ted.com/the-genius-of-frugal-innovation.

Tata Motors. 2018. *The Tata Nano Car*. Tata, Available at: https://nano.tatamotors.com.

Tiwari, R., and Herstatt, C. 2012. 'Assessing India's lead market potential for cost-effective innovations,' *Journal of Indian Business Research* (4:2), pp. 97–115.

Whittington, R., Yakis-Douglas, B., and Ahn, K. 2015. 'Cheap talk? Strategy presentations as a form of chief executive officer impression management,' *Strategic Management Journal* (37:12), pp. 2413–2424.

Williamson, P.J. 2010. 'Cost Innovation: Preparing for a 'value for money' Revolution,' Long Range Planning (43:1), pp. 343–353.

Williamson, P.J., and Zeng, M. 2009. 'Value for Money Strategies for Recessionary Times,' *Harvard Business Review*. March.

NOTES

Chapter 2
1. This case is adapted from Marabelli and Newell (2009).

Chapter 3
1. This case is adapted from Huang et al. (2015). The terms *univocality* and *multivocality* are expanded and explained fully in the case study text.

2. Ideal, not in the sense of ideal for individuals, but in the sense of most representative of this form of efficient organisation.

Chapter 4
1. This case is adapted from Huang et al. (2014).

Chapter 5
1. This case is adapted from Swan et al. (2016).

Chapter 6
1. This case is adapted from Morton et al. (2016).

Chapter 7
1. This case is adapted from Marabelli et al. (2017) and Marabelli et al. (2014).

2. The practice perspective (the practice-based view or, in this particular context, the 'practice turn') is an articulated epistemological perspective that revolves around the idea that knowledge is created through practice (i.e., doing) and is a social (and material, as we will explain later) accomplishment. However, the ontologies underpinning these epistemologies vary – see Stein et al. (2014) for a review; here, we discuss the 'practice perspective' only in general terms and rely on its sole contribution to epistemology.

Chapter 8

1. This case is adapted from Leong et al. (2016).

2. It is, however, worth noting that cutting-edge innovations such as Fitbit (a watch that captures our exercise and calculates how many calories we burn – or we should burn – on a daily basis) might change the negative view of mobile technologies with respect to personal health.

Chapter 9

1. This case is adapted from Marabelli et al. (2017).

Chapter 10

1. This case is adapted from ongoing work by Morton (2019).

INDEX

A

ABS

    Absorptive capacity, 26, 156

    innovation and, 40–44, *see* Anti-lock
braking system (ABS)

Adopters, 34, 35

Advisory committees, 165, 166, 168

Agency, 169–170, 172

    in innovation, importance of, 6–8

The Age of Discontinuity, 8

Agile methodologies, 125

Airbnb, 29, 210, 211

Algorithm-based GPS systems, 244

Algorithmic decision-making, 232–236

Algorithms process, 238–240

    impact on innovation, 243–246

Alibaba, 192, 197, 198, 201

Almeida, P., 156

Amazon, 86, 87, 150, 152–153, 205, 208,
209, 249

Ambidexterity, 91–94

Ambiguities affects, 238–240

Ambos, B., 63

Andreu, R., 10

Anti-lock braking system (ABS), 257

Article 3 of the Japanese Civil Code, 138

Artificial Intelligence, 10

Asbury Automotive Group, 4

Association for Information System's
e-library, 10

AT&T, 29

Automation, 3–4

Automotive industry, 6, 85

    sensor-based technologies in,
224–229

B

Bargaining power

    of customers, 84–86

    of suppliers, 84

Barnard, C., 56

Bayus, B.L., 207

Beishan village, 193–194

    evaluation of, 194

Berkeley Museum of Vertebrate Zoology, 178

Bessant, J., 107

Bhagwan Mahaveer Viklang Sahayata
    Samiti (BMVSS), 258–260

Big data, 232

    analytics of, 240

    comparison with little data and, 233–236

    innovation opportunities from, 246–247

Birth of knowledge management, 8–9

Black-box technology, 224, 226

Blood-work report, 178

BMW, 193

Boundary objects, 173, 177–181

    comparison with strategic and, 180–181

    types of, 178–180

Boundary spanning, 153–155

BPI, *see* Business Process Improvement

BPR, *see* Business Process Reengineering

Brenner, B., 63

Brokering, 153–155

Brown, J. S., 13–15, 42

Brynjolfsson, E., 231

Bureaucracies

    digital innovation and change, 60–61

    dysfunctions of, 55–57

Burns, T. E., 58, 59

Burt, R. S., 153

Business models and open innovation,
149–151

Business Process Improvement (BPI), 27

Business Process Reengineering (BPR), 27

Business-to-business (B2B) ticketing
    platform, 80

C

Canada-care case, 170, 173, 175, 176, 180,
184–186

    medical sheet in, 168

    role of objects in innovative healthcare
initiative in, 164–166

Capital, 2
Caramela, S., 53
Carlile, P., 179, 180
Centralised decision-making, 61
Charismatic authority, 53
Checkland, P., 14
Chesbrough, H. W., 144, 148, 149,
    151–153
China, digital innovation in, 192–194
ChinaTicketCo case, 78
    as deliberate planning process
        bargaining power of customers, 84–86
        bargaining power of suppliers, 84
        threat of established rivals, 82–83
        threat of new entrants, 83–84
        threat of substitute products or
            services, 83
    digital technology and strategy-as-
        practice, 94–97
    emergent view of strategy, 89–91
    Porter's three strategies, 86–89
    strategy-as-practice and ambidexterity,
        91–94
    strategy vs. strategising, 81–82
    ticketing practices in China
        transactional and relational practice,
            79–80
        transactional practice, 78
        transactional, relational and
            experiential practice, 80–81
Ciborra, C. U., 10
Circular economy, 271
Circular value networks, 276
Classical organisation design, 53–55
Clinton, H., 236
Cloud, 96
Codification strategy, 9
Cohen, W. M., 40, 42
Co-located programme, 108
CommCo case, 203
    bureaucracies, 60–61
    classical organisation design, 53–55
    communication culture for OPC and
        UGC, 52–53
    content on the intranet, types of, 50–52
    contingency theories, 58–59
    dysfunctions of bureaucracy, 55–57
    new forms of organising, 62–70
    organisational forms, aligning old and
        new, 70–71

Communication channels, 34, 60
Communities of practice, 16, 141
Complex project ecology, 109
Complicated and complex innovation
    development, 118–119
    projects, 117–118
Computing applications, 236–240
Consequentialism, 264
Content on the intranet, types of, 50–52
Contingency theories, 58–59
Control, 61
Cook, S. D., 13–15, 42
Corporatism, 69
Cosmetic Skin, 103
Cost–benefit analysis, 265
Courpasson, D., 175
CRM system, see Customer Relationship
    Management system
Cross-docking, 88
Cross, R., 63
Crowdsourcing, 204–209
Crutch technology, 250
Cummings, J. N., 63
Customer Relationship Management
    (CRM) system, 31

D
Data
    and digital world, 229–232
    impact on innovation, 243–246
Data-driven decision-making, 234
Datafication, 230–235
Debating the Sharing Economy, 211
Decentralised decision-making, 62–64
Decision-making process, 176, 177, 194,
    229–236, 240, 243, 245, 264
Defence-co's innovationJam, case, 132–135
    open innovation, 144–147
        business models and, 149–151
        and governance, 157–158
        inside-out and outside-in open
            innovation, 148–149
        and networking, 153–155
        search issue, 155–156
        and services, 151–153
    project organising and liminality,
        137–141
    projects and liminality, paradoxes in,
        141–143

Define, Measure, Analyse, Improve and Control (DMAIC) methodology, 27
Deliberate planning process, strategy
  bargaining power of customers, 84–86
  bargaining power of suppliers, 84
  threat of established rivals, 82–83
  threat of new entrants, 83–84
  threat of substitute products or services, 83
Dell, 88, 207
Deng, X., 206, 207
Dental Skin, 103
Desouza, K. C., 108
Development of innovative drugs, 118
Differentiation strategy, 87, 88
Diffusion of innovation
  antecedents of, 33–35
  processes and, 35–36
Diffusion of Innovation (Rogers), 25
Digi-housekeeping, 215
Digital connectivity
  implicit *vs.* explicit, 197–198
  See also Explicit digital connectivity
Digital divide, 199, 203
Digital innovation, 5–6, 36, 124, 260
  agency in innovation, importance of, 6–8
  in china, 192–194
  cycle, 6
  future of
    challenges of frugal innovation, 273–275
    developing and developed economies, 268–269
    frugal innovation in india, 256–257
    innovation responsibility, 262–266
    Jaipur Foot, 258–260
    key actions for successful responsible and frugal innovation, 275–277
    key opportunities and challenges for responsible, 266–268
    responsible and frugal innovation, 278–280
    stablished models of innovation, 269–272
    Tata Nano, 257–258
  making of, 247–249
  objects in knowledge sharing and, 172–173
  work–life balance and, 215

Digital technologies, 5, 9, 69, 106
  and strategy-as-practice, 94–97
Digital tool, 156
Digital world, data and, 229–232
Digitised objects, 241–246
Disruptive innovations, 29, 58
Distributed project, 108
DMAIC methodology, *see* Define, Measure, Analyse, Improve and Control methodology
'Double-loop' learning processes, 116
Dougherty, D., 116, 117
Dreyfus, H., 186
Dropbox, 31
Drucker, P., 8, 149
Dunne, D. D., 116, 117
Dysfunctions of bureaucracy, 55–57

E
Earl, M. J., 154
eBay, 198, 201
E-commerce, 198–201, 203
  ecosystem, 195
  platform, 200
Edelman, B.G., 212
Edmondson, A. C., 109
Effective project management, 114–116
EHR systems, *see* Electronic health record systems
Electronic health record (EHR) systems, 170
Electronic medical record (EMR) system, 166, 170
Emergency Room (ER), 166, 167
Emergent process, 82
Emergent view of strategy, 89–91
Employee tracking systems, 244
EMR system, *see* Electronic medical record system
End-of-Life Vehicles Directive in 2000, 6
Engwall, M., 111
Enterprise social media (ESM) systems, 202–204
Enterprise System (ES)
  failure (2001–2003), 23–24
  in organisations, 25
  story, 22–23
  success (2004–2010), 24–25
Epistemology of practice, 14
ER, *see* Emergency Room

ES, *see* Enterprise System
European Union's Horizon 2020, 266
Evaristo, J. R., 108
Everything Bonsai!, 208
Explicit digital connectivity, 198–201
    and crowdsourcing, 204–209
    and use of social media within
      organisations, 201–203
    vs. implicit digital connectivity,
      197–198
    and work–life boundaries, 212–215
Explicit knowledge, 11–12, 196
Exploitation, knowledge exploration *vs.*,
    9–11
External knowledge, 41

F
FedEx, 27
Feldman, M. S., 107
Felin, T., 146, 157
2008 financial crisis, 238
Firm/user perspectives of innovation, 8, 10,
    32, 87
Flatter organisations, 63, 64
Fleming, L., 154
Ford, H., 5
Ford Pinto case, 265
Fostering innovation, role of knowledge
    in, 40
'Four Lessons Amazon Learned From
    Webvan's Flop' (Forbes), 86
Freeman, J., 62
Frugal innovation, 268–269
    challenges of, 273–275
    concept and responsible for,
      278–280
    and established models of innovation,
      269–272
    in india, 256–257
    key actions for successful responsible
      and, 275–277
    principles for, 271–272
    responsibility of, 262–266, 278–280

G
Gantt charts, 120, 124, 174
GAP, *see* Guaranteed Asset Protection
Garsten, C., 139, 141
Gates, B., 94

Generali Group, 225
Generali Italia (EU), 225–227
Generative dance, 15
Genetically modified (GM) crops, 84
George, G., 41
GFT, *see* Google Flu Trend
Global markets, 4
Global positioning system (GPS), 224, 225,
    227, 229, 244
GM crops, *see* Genetically modified crops
GNU, 209
GNU-licenced software movements, 210
Google, 172
Google Flu Trend (GFT), 237
Google search engine, 237
Gouldner, A., 56
Governance model, 157
Governance, open innovation and,
    157–158
GPL, 209
GPS, *see* Global positioning system
Grabher, G., 108
Grassroots Association, 201
Griesemer, J.R., 177, 178, 183
Grouping, 110
gThrive, 271
Guaranteed Asset Protection (GAP), 4
Guardian article, 209

H
Hagen, R., 140
Hammer, M., 27
Hannan, M. T., 62
Hansen, M., 9
Hardy, C., 176
Harvard Business Review, 174
Healthcare, 278–279
Hierarchical *vs.* 'flat' organisations, 62
Huang, J., 70
Human agency, 173
Human-centric nature of innovation, 7
Human engagement, in objects process,
    185–186
Hu, N., 208

I
'Ideal' bureaucratic form of organising,
    54–55
IdeaStorm, 207

Immanent, 14
Implicit digital connectivity
    explicit digital connectivity *vs.*,
    197–198
    opportunities and challenges for
    innovation related to
        algorithmic decision-making and
        big, little data, 232–236
        and computing applications,
        236–240
        data and digital world, 229–232
        and digitised objects, 241–246
        innovation opportunities from big
        and little data, 246–247
        making of digital innovation,
        247–249
        sensors in automotive industry,
        224–229
Industrial Revolution, 2
Industrial symbiosis, 6
Informal communication, 56
Information Systems (IS), 95
    literature, 10
Information Technology (IT), 3–4
Initiative process, objects in, 166–169
Innovation, 4–5, 81
    and absorptive capacity, 40–44
    development of, 5
    digital innovation, growing
    importance of, 5–6
        agency in innovation, importance
        of, 6–8
    Enterprise System (ES)
        failure (2001–2003), 23–24
        in organisations, 25
        story, 22–23
        success (2004–2010), 24–25
    firm/user perspectives of, 32
    incremental, radical and disruptive,
    28–30
    interactive view of, 37–39
    liminality, 139
    power and knowledge absorption,
    44–45
    product and process, 26–27
    Rogers' diffusion of, 25–26
    Rogers' model of the diffusion of,
    33–37
    routines and, 105–107
    service innovation, 30–31

    types of, 30
Innovation process, 171–173
Innovation projects, uncertainty in
    market uncertainties, 122
    organisational uncertainties, 123–124
    resource uncertainties, 122–123
    technical uncertainties, 121–122
Innovative drugs, development of, 118
Innovative power of objects, 181–185
An Inquiry into the Nature and Causes of the
    Wealth of Nations (Smith), 55
Inside-out and outside-in open innovation,
    148–149
Institutional context, projects in, 124–126
Integrative capabilities, 69
Intellectual Property (IP), 121
Interactive innovation, 116
Interactive process, 42
Interactive view of absorptive capacity, 42–44
Internet-enabled computing technologies,
    192
Internet of things, 229
Intranet, 52
IP, *see* Intellectual Property
IS, *see* Information Systems
IT, *see* Information Technology

J
Jaipur Foot, 258–260
Jamming activity, 133
Jarzabkowski, P., 177
Jobs, S., 29, 64
Johnson, P. E., 13
Jugaad, 269

K
Kaplan, S., 177
Key performance indicators (KPIs), 277
KM, *see* Knowledge management
KMS, *see* Knowledge Management Systems
Knowing, 14
Knowledge, 11–12, 88
    ability to discrimination, 12–13
Knowledge absorption, power and, 44–45
Knowledge and innovation, implications
    for, 9–11
        birth of knowledge management, 8–9

knowledge exploration *vs.* exploitation, 9–11
digital innovation, growing importance of, 5–6
  agency in innovation, importance of, 6–8
  knowledge economy, 2–3
    automation, 3–4
    services, importance of, 4–5
Knowledge-based view, 196
  of firm, 8
Knowledge economy, 2–3, 8
  automation, 3–4
  services, importance of, 4–5
Knowledge exploration *vs.* exploitation, 9–11, 41, 92
Knowledge gap, 7
Knowledge management (KM)
  birth of, 8–9
    knowledge exploration *vs.* exploitation, 9–11
  in organisations, 11
Knowledge Management Systems (KMS), 9
Knowledge sharing, 10
  in crowd activity, 207–208
Kodak, 85–86, 144, 146
KPIs, *see* Key performance indicators

L
Labour, 2, 3, 8
Land, 2
Lane, P. J., 41
Lave, J., 16
Learning boundary, 111
Legal-rational authority, 53
Legitimate peripheral participation, 16
Lego, 152
Leonardi, P.M., 202
Levinthal, D. A., 40, 42
Liminality, 108
  project organising and, 137–141
  and projects, paradoxes in, 141–143
Liminal space, 137
  characteristics of, 139–141
Lindgren, M., 115
Linear and interactive views of innovation, 37–39
Linux operating system, 210
Lipase inhibitor, 122–123

Little data, 232
  analytics of, 240
  comparison with big data and, 233–236
  innovation opportunities from, 246–247
Little, J., 213
Long-term ROI, 177
'Longwall' method, 171
Low cost strategy, 87
Lubatkin, M., 41
Luca, M., 212

M
MAC 400, 272
Majchrzak, A., 207
Malhotra, A., 207
'The Management Theory of Max Weber', 53
Market uncertainties, 122
Marra, G., 239
Material agency, 173
Materiality role, 173
Matrix structures, 67
McAfee, A., 231
McKinsey, 278
Mechanical Turk (MTurk), 205–206
Mechanistic organisations, 59
Merton, R., 56
Meulman, F., 156
Meyer, J. W., 125
Mintzberg, H., 58, 81–82, 90
Mitticool fridge, 273, 274
Mobile working model, 51
Modern laser disk (CD) reader, 30
Modularisation, 65, 66
Monsanto, 84
Mr Lv, 193, 194
MTurk, 249
Multivocal environment, 68

N
National Innovation Foundation, 258
National Institute of Clinical Excellence (NICE), 122
Nelson, R. R., 106
Nested nature of projects, 124–126
Network approach, 196

Networking, open innovation and, 153–155

New business models, 5

Newell, S., 108

New ideas, development and implementation of, 7

New organisational processes, 5

New products, 5

New services, 5

New York Times article, 239–240

NGO, *see* Non-governmental organisations

NICE, *see* National Institute of Clinical Excellence

Nicolini, D., 91

Non-consequentialism, 264

Non-digital innovation, 260

Non-governmental organisations (NGO), 192

O

OBD system, *see* Onboard diagnostics system

Objects process, 169–173

    human engagement in, 185–186

    in initiative process, 166–169

    innovative power of, 181–185

    in knowledge sharing, 172–173

    power and role of, 184

Office of the Comptroller of the Currency, 266

Onboard diagnostics (OBD) system, 224–227, 229, 232, 234

Online channel, 80

Online Shop Association, 192, 193, 201

OPC, *see* Organisation Published Content

OpenIDEO, 207

Open innovation, 144–147

    business models and, 149–151

    and governance, 157–158

    inside-out and outside-in open innovation, 148–149

    and networking, 153–155

    search issue, 155–156

    and services, 151–153

Open Innovation: The New Imperative for Creating and Profiting from Technology (Chesbrough), 144

Openness, 52

Orchestrated approach, 200, 202

Organic approaches, 200, 202

Organic organisations, 59

Organisational forms, 70–71

Organisational learning, 43

Organisationally controlled crowd involvement, 204–205, 212

Organisational uncertainties, 123–124

Organisation, digital communication, 51

Organisation Published Content (OPC), 51, 52

    communication culture for, 52–53

Organisations, 62, 139

    Enterprise system in, 25

    knowledge management in, 11

Organising for digital innovation

    bureaucracies, 60–61

    classical organisation design, 53–55

    CommCo, 50

    communication culture for OPC and UGC, 52–53

    content on the intranet, types of, 50–52

    contingency theories, 58–59

    dysfunctions of bureaucracy, 55–57

    new forms of organising, 62–70

    organisational forms, aligning old and new, 70–71

Organising, new forms of

    business units, 65–67

    flatter structures and decentralised decision-making, 62–64

    networking, 69–70

    project forms, 67–68

Original 'Skin' product, 103

Orlikowski, W.J., 172

Orr, J. E., 16

Ostensive aspect of routine, 107

Owen-Smith, J., 69

P

Padula, G., 70

PageRank, 172

Palo Alto Research Centre (PARC), 149

Paper-based system, 167, 169

PARC, *see* Palo Alto Research Centre

Pariser, E., 239

Pay as you drive (PAYD) programmes, 224, 225, 227–228

PAYD programmes, *see* Pay as you drive programmes

Peer-to-peer crowd involvement, 209–212
Penske Automotive Group, 4
Pentland, B. T., 107
Performative aspect of routine, 107
Personalisation strategy, 10
PERT, *see* Programme evaluation and
  review technique
PERT diagrams, 174
Pharma, 102
Pilot project, 164, 165, 170
Pisano, G., 174
PMBOK, *see* Project management body of
  knowledge
Polanyi, M., 11
Population ecology, 62
Porter, M., 81, 86, 153
Porter's five forces model
  bargaining power of customers, 84–86
  bargaining power of suppliers, 84
  threat of established rivals, 82–83
  threat of new entrants, 83–84
  threat of substitute products or
    services, 83
Porter's three strategies, 86–89
Possession, 13
  perspective focus, 13–14
  and practice perspective, 14–16
Power and knowledge absorption, 44–45
Power dynamics, 67
PowerPoint presentations, 166, 173–176,
  180, 185
Prabhu, J., 275–276
Practice view of knowledge, 16
Practised knowledge, 13–16
Pragmatic boundary, 180
Praxis, 91–93
Process-based capabilities, 155
Process innovation, 26–27
Process view of absorptive capacity, 41–42
Product innovation, 26–27
Productivity paradox, 3
Programme evaluation and review
  technique (PERT), 113–114
Progressive Insurance (USA), 227–229
Progressive policyholders, 228–229
Project learning, uncertainty in, 119–121
  market uncertainties, 122
  organisational uncertainties, 123–124

resource uncertainties, 122–123
technical uncertainties, 121–122
Project management, 110–112, 116–117
  characteristics, 112–114
  complicated and complex innovation
    development, 118–119
  complicated and complex innovation
    projects, 117–118
  effective project management,
    114–116
Project management body of knowledge
  (PMBOK), 112–113, 115
The Project Management Body of Knowledge,
  112
Project Management Institute, 112
Project organising and liminality, 137–141
Projects
  and liminality, paradoxes in, 141–143
  nested nature of, 124–126
  overview of, 108–109
  and project management (*see* Project
    management)
  and teams, 109–110
  uncertainty in, 119–121
    market uncertainties, 122
    organisational uncertainties,
      123–124
    resource uncertainties, 122–123
    technical uncertainties, 121–122

Q
Qarnot Computing, 271

R
Radical innovations, 29
Radio-frequency identification (RFID), 224
Radjou, N., 270, 275–276
R&D, 270, 271, 274
  department, 174
Real-time visibility, 279
'Reduce, reuse and recycle', mantra of, 6
Relational capabilities, 69
Repository approach, 196
Resource uncertainties, 122–123
Responsible innovation, 257, 258,
  261–268, 272, 277, 279
  challenges in, 267–268

Retail stores, 279
RFID, *see* Radio-frequency identification
Rice, M. P., 119–121, 125
Rivalry, 82–83
Rogers' diffusion of innovation, 25–26
Rogers' model of the diffusion of
    innovation, 33–37
ROI, 175
Rosenkopf, L., 70
Routines, 90
    and innovation, 105–107
Rowan, B., 125
Royer, I., 148

S

Scaffolding, 106
Scientific Management, 2–3, 115
Scripts, 90–91
Seattle Foot, 259
Security analytics, 279
Seeds of Death (Monsanto), 84
Selznick, p., 56
Semantic boundary, 180
Sense of community, 141
Sensor-based technologies, in automotive
    industry, 224–229
Service innovation, 30–31
Services
    importance of, 4–5
    open innovation and, 151–153
Shaw, M.L., 62, 63
Short-term ROI, 24, 177
'Shortwall' method, 171
Silicon Valley, 270, 271
SIM, *see* Subscriber identification module
Simple modularisation, 66
Sinclair, C., 28, 36
Singular 'best practice' organisational
    design, 53
Site of knowing, 91
Six Sigma system, 27
SkinTech, case, 102–105
    projects and project management,
        108–112, 116–117
        characteristics, 112–114
        complicated and complex
            innovation development,
            118–119

complicated and complex
    innovation projects, 117–118
effective project management,
    114–116
nested nature of, 124–126
routines and innovation, 105–107
uncertainty in projects and project
    learning, 119–121
    market uncertainties, 122
    organisational uncertainties,
        123–124
    resource uncertainties, 122–123
    technical uncertainties, 121–122
Skyrme, D. J., 154
Small firms, 119
Smith, A., 55
SMS Medical College Hospital, 259
Snapshot programme, 227, 228
Social software, 198
Social system, 35
Society of Indian Automobile
    Manufacturers, 258
SOE, *see* State-owned enterprises
Software innovation, challenges of, 125
Solid ankle cushioned heel (SACH), 259
Spinning Jenny, 2
S-shaped curve, 35–36
Stalker, G. M., 58, 59
Star, S.L., 177, 178, 183
State-owned enterprises (SOE), 78
Steering committees, 165
Stevens, B., 5
'Stock' of internal knowledge, 40
Strategic maps, 174
Strategic objects, 172, 174–177, 181
    alternative views on, 175–176
    effectiveness of, 176–177
Strategising for digital innovation
    ChinaTicketCo, 78
    as deliberate planning process
        bargaining power of customers,
            84–86
        bargaining power of suppliers, 84
        threat of established rivals, 82–83
        threat of new entrants, 83–84
        threat of substitute products or
            services, 83
Strategising for digital innovation (*cont.*)
    digital technology and strategy-as-
        practice, 94–97

emergent view of strategy, 89–91
Porter's three strategies, 86–89
strategy-as-practice and ambidexterity, 91–94
strategy *vs.* strategising, 81–82
ticketing practices in China
   transactional and relational practice, 79–80
   transactional practice, 78
   transactional, relational and experiential practice, 80–81
Strategy
   conception and execution of, 87
   practitioners, 90–91
   vs. strategising, 81–82
Strategy-as-practice, 90, 91–94
   approach, 174, 181
   digital technology and, 94–97
Structural hole, 153
Subscriber identification module (SIM), 225
Success, principles for, 109–110
Suichang village, 192–193, 200–202
   evaluation of, 194
Supply chain management, 10
'The Surprising Ways Car Dealers Make The Most Money Off You,' 4
Swan, J., 111
Switching costs, 86
Syntactic boundary, 180
Systematic soldiering, 2

T
Tacit knowledge, 12, 196
Taobao villages, 192–194, 199, 200
Tata Group, 257, 269
Tata Motors, 257
Tata Nano, 256, 257–258
Tata, R., 257
Tata Swach, 258
Tavistock Institute, 171
Taylor, F. W., 2, 3, 243
Team leadership, 110
Teams, projects and, 109
Tech Co.
   failure (2001–2003), 23–24
   in organisations, 25
   story, 22–23
   success (2004–2010), 24–25

Technical uncertainties, 121–122
Telematic service provider (TPS), 225
Tennessee Valley Authority (TVA), 50–57
Thomas, J., 112, 114–116
Threat intelligence network, 279
Ticketing practices in China
   transactional and relational practice, 79–80
   transactional practice, 78
   transactional, relational and experiential practice, 80–81
Time magazine, 260
Townley, B., 115
TPS, *see* Telematic service provider
'Traditional' approach, 32
Traditional authority, 53
Traditional factors of production, 2
Transport system, 279
TripAdvisor, 208
Tsoukas, H., 13
Turner, V., 140
TVA, *see* Tennessee Valley Authority

U
Uber, 198, 210, 211
Ubiquitous computing, 197, 229
UGC, *see* User Generated Content
Uncertainty in projects and project learning, 119–121
   market uncertainties, 122
   organisational uncertainties, 123–124
   resource uncertainties, 122–123
   technical uncertainties, 121–122
Uninformed control, 239
United Parcel Service (UPS), 153
Univocal communication, 68
Unknowability, 117, 119
UPS, *see* United Parcel Service
USA, 259
User-controlled crowd involvement, 208–209, 212
User Generated Content (UGC), 51, 52
   communication culture for, 52–53

V
Van de Ven, A. H., 7, 13
van Gennep, A., 137
Vasudeva, G., 69

Vienna's co-creation project, case, 135–136
    open innovation, 144–147
      business models and, 149–151
      and governance, 157–158
      inside-out and outside-in open
        innovation, 148–149
      and networking, 153–155
      search issue, 155–156
      and services, 151–153
    project organising and liminality,
      137–141
    projects and liminality, paradoxes in,
      141–143
Virtual Learning Environment (VLE), 245
Vladimirou, E., 13
VLE, *see* Virtual Learning Environment
von Hippel, E., 204
von Krogh, G., 207

W
Wagner, E. L., 143
Waguespack, D., 154
'Walking data generators,' 231, 232, 247

Walmart, 88
Weber, M., 53, 54
Webvan, 86
WeCash, 36
Wenger, E., 16, 141
Whittington, R., 90
Wikipedia, 182, 184
Winter, S. G., 106
Wong, Faye, 79
Woodward, J., 58
Work–life balance, 215
Work–life boundaries
    changing nature of, 213–215
    explicit digital connectivity and,
    212–215

Z
Zahra, S. A., 41
Zenger, T., 146, 157
Zhejiang Province of China, 192
Zhou, Y. M., 67
Zipcar, 211